# Milady SalonOvations' Public Relations for the Salon

## Online Services

**Delmar Online**
To access a wide variety of Delmar products and services on the World Wide Web, point your browser to:
  http://www.delmar.com
  or email: info@delmar.com

**thomson.com**
To access International Thomson Publishing's home site for information on more than 34 publishers and 20,000 products, point your browser to:
  http://www.thomson.com
  or email: findit@kiosk.thomson.com

A service of I(T)P®

# SalonOvations' Public Relations for the Salon

by
Jayne Morehouse

Milady Publishing
(a division of Delmar Publishers)
3 Columbia Circle, P.O. Box 12519
Albany, New York 12212-2519

## NOTICE TO THE READER

Publisher does not warrant or guarantee any of the products described herein or perform any independent analysis in connection with any of the product information contained herein. Publisher does not assume, and expressly disclaims, any obligation to obtain and include information other than that provided to it by the manufacturer.

The reader is expressly warned to consider and adopt all safety precautions that might be indicated by the activities herein and to avoid all potential hazards. By following the instructions contained herein, the reader willingly assumes all risks in connections with such instructions.

The publisher makes no representation or warranties of any kind, including but not limited to, the warranties of fitness for particular purpose or merchantability, nor are any such representations implied with respect to the material set forth herein, and the publisher takes no responsibility with respect to such material. The publisher shall not be liable for any special, consequential, or exemplary damages resulting, in whole or part, from the readers' use of, or reliance upon, this material.

---

Cover Design: Spiral Design Studio

**Milady Staff**
Publisher: Gordon Miller
Acquisitions Editor: Marlene McHugh Pratt
Project Editor: Annette Downs Danaher
Production Manager: Brian Yacur
Production and Art/Design Coordinator: Suzanne Nelson

COPYRIGHT © 1997
Milady Publishing
(a division of Delmar Publishers)
*an International Thomson Publishing company*

Printed in the United States of America
Printed and distributed simultaneously in Canada

**For more information, contact:**
SalonOvations
Milady Publishing
3 Columbia Circle , Box 12519
Albany, New York   12212-2519

All rights reserved. No part of this work covered by the copyright hereon may be reproduced or used in any form or by any means—graphic, electronic, or mechanical, including photocopying, recording, taping, or information storage and retrieval systems—without the written permission of the publisher.

1  2  3  4  5  6  7  8  9  10   XXX   01  00  99  98  97  96

**Library of Congress Cataloging-in-Publication Data**

Morehouse, Jayne.
     SalonOvations' public relations for the salon / by Jayne Morehouse.
       p.     cm.
     Includes index.
     ISBN: 1-56253-271-5
     1. Beauty shops—Management.   2. Beauty shops—Marketing.   3. Public relations.
I. SalonOvations' (Firm)   II. Title.
TT965.M67   1996                                                                                        96-11103
659.2'646724–dc20                                                                                          CIP

# Table of Contents

About the Author ..................................................................viii
Acknowledgments ..................................................................ix

Introduction Hey...I Saw You in the Newspaper Yesterday!....1

Chapter 1    What Is Public Relations? .......................................7
   A Business of Relationships..................................................7
      Consistency Counts .........................................................8
      What Public Relations Can't Do for You .........................8
   Some Terms to Know .........................................................10
   Dos and Don'ts of Successful PR..........................................15

Chapter 2 Honing Your Salon's Image ......................................19
   What is your Salon's Image?...............................................19
   The Hairmeister of Las Olas is
      on Call to the Press 24 Hours a Day! ..............................20
   "What's My Image" Worksheet ..........................................22
   Collateral Materials .............................................................25
   A Word About Photo Shoots...............................................30
      Model and Photographer Releases .................................32
   Your First Press Mailing: The Backgrounder.......................35
      Press Releases ................................................................37
   Building Relationships ........................................................40

Chapter 3    Planning Your PR Program ...................................43
   What Do You Want Out of PR? ..........................................43
      How to Score Your Answers ..........................................44
   Developing Your PR Program .............................................44
      Targeting Your Message.................................................45
   A PR Plan Example .............................................................50

  Outlining Your Plan ................................................................55
    Where to Send Your Message ........................................57

**Chapter 4 Hiring and Budgeting for PR Pros** ......................59
  Do You Need to Work with a PR Pro?..................................59
    Screening Potential PR Agents ......................................61
    Making Choices................................................................63
  Budgeting................................................................................64

**Chapter 5 Creating a Name in Your Own Backyard** ..........73
  Your Event Checklist and Budget Form ..............................75
  Why Local PR?........................................................................75
    Beginning a Local PR Program ....................................76

**Chapter 6 Making a Name Nationally**..................................85
  Meet Lenny LaCour................................................................87
  National Publications: What Editors Want ........................89
    *Good Housekeeping* ......................................................89
    *Self* ..................................................................................89
    *Marie Claire* ..................................................................90
    *Redbook*..........................................................................90
  Hairstyling Publications: What Editors Want ....................91
    Harris Publications ........................................................91
    GCR Publications ..........................................................92
    *Sophisticate's Hairstyle Guide* and
    *Sophisticate's Black Hairstyles & Care Guide* ............93
  National Television ................................................................94

**Chapter 7 Getting Known in Your Industry** ........................97
  Trade Publications: What Editors Want ..............................99
    *Modern Salon*, East and West Coasts ..........................99
    *Salon News* ..................................................................103
    *American Salon* ............................................................106

　　　　*SalonOvations* ................................................................... 109
　　　　*Nails* Magazine ................................................................ 112
　　　　*Passion* and *Coiffure Q* .................................................. 115
　　　Getting a Bigger PR Bang .................................................. 117
　　　Join a Group; Get More PR ............................................... 118

Chapter 8　Special Events, Charity Work
　　　　　　and Other Networking Opportunities ............... 121
　　　Getting Behind a Good Cause ........................................... 123

Chapter 9　Where Do I Go From Here? ................................ 129
　　　Your Image Starts and Stops Here ..................................... 129
　　　PR Ideas From A to Z ........................................................ 131

# Appendices
　　　Appendix 1 Local Media List ............................................ 143
　　　Appendix 2 Popular Consumer Magazines ...................... 149
　　　Appendix 3 Consumer Hairstyling Publications .............. 155
　　　Appendix 4 Trade Magazines ........................................... 157
　　　Appendix 5 PR Activity Log .............................................. 163
　　　Appendix 6 Press Clippings Record .................................. 165
　　　Appendix 7 Press Contacts ............................................... 167

Glossary/Index ................................................................... 169

# ABOUT THE AUTHOR

## Jayne Morehouse

Jayne Morehouse is president of Morehouse Communications, Inc., a leading public relations and marketing services agency serving the beauty industry. A beauty and health writer for more than 10 years and the former editor of *American Salon* magazine, she knows firsthand what the media wants—and doesn't want. She has parlayed that knowledge into strategies and programs that make sure her company's clients, including beauty product manufacturers, educators, and beauty salons, are seen in all the right places: trade and consumer magazines, newspapers, radio, television, and even the 'Net. Jayne has pioneered a unique "relationship approach" to public relations. She and her staff maintain consistent contact with hundreds of trade and consumer editors, producers, and writers across the country and as a result, her company's clients regularly appear in the national media as prestigious as *Vogue, Elle, W, Allure, Glamour, Mademoiselle, Self, Redbook, Family Circle, Woman's Day, Modern Salon, Salon News, SalonOvations, American Salon, Nails, Nail Pro, Passion, Coiffure Q*, and more. Jayne is also coauthor of *Quick Fixes for a Bad Hair Day*.

# Acknowledgments

This book could not have been written without the sharing of personal experiences and information from many. Milady Publishing Company gratefully acknowledges the following individuals who have contributed their expertise and experience:

Frank Alvarez, Mark Frank Hair Salons, Cleveland, OH

Beauty and Barber Supply Institute for the North American Hairstyling Awards

Julie Becker, National Cosmetology Association, St. Louis, MO

Melissa Bedolis, Editor in Chief, *Salon News*, New York, NY

Suzanne Bersch, Suzanne Bersch Inc., New York, NY

Jesse and Flo Briggs, Yellow Strawberry Global Salons, Ft. Lauderdale, FL

Richard Calcasola, Maximus Salon and Spa, Merrick, NY

Noel de Caprio, Nöelle Spa for Beauty & Wellness, Stamford, CT

Cyndy Drummey, Editor in Chief, *Nails Magazine*, Redondo Beach, CA

Karyn Repinski, Beauty Editor, *Good Housekeeping*

Sharon Esche and Alex Irving, Esche and Alexander Public Relations

Edwin Fontanez, Dino Palmieri Salon, Cleveland, OH

Catherine Frangie, Milday Publishing, Albany, NY

Mary Greenberg, Editor, Harris Publications, New York, NY

Judy Guerin, New York, NY

Leslie M. Hahn, LMH Public Relations, CT

Rhonda Hicks, E'Mages by Hair Station, Houston, TX

Leslie Huntley, International Beauty Shows, Cleveland, OH

Barbara Jewett, Managing Editor, *SalonOvations*, Albany, NY

Lorraine Korman, Editor in Chief, *American Salon*, New York, NY

Sandra Kosherick, Managing Editor, GCR Publications, New York, NY

Bonnie Krueger, Editor in Chief, *Sophisticate's Hairstyle Guides*, Chicago, IL

Lenny LaCour, Spa de LaCour, Chicago, IL

Robert LaMorte, Robert Jeffrey Hair Studios, Chicago, IL

Norma A. Lee, New York, NY

Maureen Meltzer McGrath, Beauty News Editor, *Self*, New York, NY

Marcia Menter, Beauty and Style Editor, *Redbook*, New York, NY

Lisa Miller, Decatur, IL

Beth and Carmine Minardi, Minardi Salon, New York, NY

Helen Moy, Editor, *Passion* and *Coiffure Q*

Maggie Mulhern, Beauty Editor, *Modern Salon*, New York, NY

Suzanne Munshower, Suzanne Munshower Communications, Burbank, CA

Alexandra Parness, Beauty and Fitness Director, *Marie Claire*, New York, NY

Victoria Spedale, Victoria Communications, Islip, NY

Jackie Summers, Editor in Chief, *Modern Salon*, Lincolnshire, IL

Doris Thomas, Hendersonville, TN

Mario Tricoci, Mario Tricoci Hair Salons and Day Spas, Chicago, IL

Victoria Wurdinger, Lomo International, Brooklyn, NY and author of *Picture This ... Your Image in Print*

# Introduction

# Hey ... I Saw You in the Newspaper Yesterday!

That's what everyone will be saying to you when you implement a public relations (PR) program that's designed to spread the news and tell a special story about you and your salon!

Did you ever wonder how the salon down the street gets to do the hair for the fashion layouts in the local newspaper? Or why its hairdressers are continuously quoted in the local beauty pages or even in *Glamour* and *Mademoiselle* magazines? Or why that salon owner appears on makeover segments on your local television station's morning show several times a year? Are you just a little bit curious about why the local radio station did a remote broadcast from that salon as part of the business's 10-year anniversary? That's successful PR at work! That salon's work might not be any better than yours. It might not even be as good. The difference is that your competitor has an effective PR program in gear to keep the editors, writers, and segment producers informed about every important move the salon makes.

Like many of us, you might think that the various media tend to seek out their own sources. But all of these media professionals are way too busy to go out on their own and research all the salons—maybe hundreds—in your town, and then select the best to feature. They rely on the salons themselves to tell them what they are doing. They also rely on their own friends and coworkers to tell them which salons they like and why they get their hair done there. Both of those are a form of PR.

Sometimes they even rely on the salons to give them ideas for their fashion columns, articles, and shows. Again, that's PR. Salon owners who constantly feed the media ideas in an informational, educational, and helpful way—as opposed to being pushy—are rewarded by being featured in print and on air.

Okay, so what is a PR program? It's the process by which you generate editorial and good will throughout your community through your local media and activities and to other target audiences, such as within the professional salon industry itself through the trade press. Forms of PR include everything from sending out press kits to making personal contacts with the media to participating in community charity work to ensuring that every client who walks down the street wearing your services on a daily basis looks fabulous.

Once implemented, over time, a good PR program will build your credibility by using local newspapers and magazines, perhaps national magazines, and also industry publications, to tell your target audience what makes you special. It will make you "known" for something...for example, as the place for trendy hair color, as the salon that cuts the best bob in town, as the salon that does all of the hair for society functions, as the most altruistic salon that's always willing to give time to help charity events and needy people, as the salon that's the best place to work within 100 miles, or whatever else you want to be known as. A personally tailored PR program lets you create and enhance whatever image you want to present, as long as your salon can back it up. (In Chapter 2, we'll explore ways to define your image to ensure that you're telling a consistent, verifiable story through your PR.)

Although this might sound a bit overwhelming, don't let it scare you. You probably already have many of the elements in place that are needed to launch an effective PR program. You just have to organize them into a productive plan, and this book will help you do that.

In fact, most salons have newsworthy activities going on. You're giving tips and trend information to your clients every day. You're probably participating in a variety of charity events from fashion shows to cut-a-thons either by yourself or in conjunction with your distributors and manufacturers. You attend beauty shows and advanced education classes. You're hiring new staff members and adding new services. All of these topics are press worthy. You just might not be tooting your own horn enough or to the right people to get the press you deserve. When properly presented, that same kind of information can get you in the news...and even on television. When a newspaper or magazine writes about you, it's a highly credible endorsement. In general, people regard an article with more respect than they do an advertisement in the very same publication. Why? A third party endorsement—someone else speaking well of you—is much stronger than saying good things about yourself. That is not to say that you shouldn't advertise. It is just a different method for creating an important message that fits within your salon's overall image. (In Chapter 1, we'll discuss the difference between PR and advertising.)

**Fig. 0-1** *Chicago salon owner Lenny LaCour knows that hobnobbing with a who's who of local and national celebrities and political officials can help him be seen in all the right places. Here, he's with Leanza Cornett, whose hair he styled during her reign as Miss America.*

More and more salon owners are realizing that PR is good for business (Fig. 0-1). Some might even say it's an essential element in their overall marketing plan. Chances are you do too, or you wouldn't have picked up this book.

What's good news is that effective PR need not be extravagant or expensive. After all, it is simply putting forth information that is interesting, easy to understand, creatively packaged to attract attention, and presented in a useful form that editors can use. PR can make all the difference in the world between being known for your expertise and what you do best, and hiding your own shooting star behind a cloud. Although PR is not free, it is cost effective. Because it carries with it the power of editorial credibility, every single PR mention of your salon in print is worth its weight in gold.

Some potential results of launching a PR campaign for your salon are obvious. Increased exposure will enhance your reputation. You can turn that into more clients, leading to greater service and retail sales, as well as the opportunity to attract even more clients by networking

**Fig. 0-2** *Strong PR programs helped these winners of the 1996 International Beauty Show's Editor's Choice Beauty Press Awards gain recognition from top media. Winners are selected annually by the top beauty and fashion press. From left: Orlando Pita, Ruth Roche, Trevor Sorbie, Suki Duggan, Chenzo Balsamo, Paula Gilmore, Mike Karg, Vivienne Mackinder, and John Sahag.*

through these new clients. A well-placed PR mention can excite your current clients. They'll tell their friends, "That's my salon in the newspaper today!" People love to feel proud of those organizations they choose to do business with! It's an affirmation that they made a good choice. The same goes for your staff. They're proud to work at a business that's featured in the news, and they're less likely to leave a job and salon they feel good about.

Of course, PR has other benefits. It can create community credibility, good will, and good feelings for your business—both within your community and within your salon itself. For example, doing the hair for your community's major fund-raiser fashion show can garner tons of good will and put your business in a great light and in important company with the leaders of the community.

In addition, PR can be a great staff motivator and morale booster for your entire business. Think, for example, how wonderful it would feel to get a great write-up because your salon helped collect food for the local homeless shelter or because your salon organized a breast cancer awareness and education program for your clients, staff, and community. What's an even bigger benefit is that with programs like that, you can also invite your clients to get involved with the planning and implementation, which makes them an even closer member of your salon's team. When they feel they're an important part of your business beyond just sitting in your chair, they are much less likely to leave, no matter what enticements the salon down the street can offer.

This book will answer common questions salon owners have about PR (Fig. 0-2). In addition, it will:

- Help you set PR goals.
- Guide you through the process of putting together a plan that will achieve the results you're after.
- Help you decide if you should hire a PR professional to spread your message or hire a professional writer, then handle the mailing and other legwork yourself.
- Give you examples and case studies of salon owners just like you who have had success with PR.
- Share what top consumer and trade magazine editors want to know from you, and how to go about approaching them.
- Share Pro's Tips, which let you in on the secrets that professional publicists know and use to position their clients in front of the media.
- Give you lists of press contacts and tips for list management.
- Share with you our professional tactics for getting an extra benefit from your PR success by using it for internal marketing within your salon. After all, think how proud your current clients will be of "their salon" when they see you quoted in *Glamour* or the local newspaper. And think how proud your staff will be of being part of such a team!

Sound like a good idea? Then let's get started!

# Chapter 1

# What Is Public Relations?

*B*eing in the image business, you already know the value of building a positive image for your clients. When launching a PR program or campaign, you are doing the same thing for yourself and your salon.

So just what is *Public Relations*? PR is the art and strategy of shaping the public's opinion about you and your business. It is a way of publicly defining who and what you are and what you stand for. PR is a huge umbrella that covers many areas and activities. It is just one piece of an overall marketing mix that also includes advertising, merchandising, and promotions.

PR is not advertising. Unfortunately, the two terms are often lumped together. PR is an excellent way to support advertising. To oversimplify the difference, advertising is "paid publicity," whereas PR is "free editorial." Which would you rather have? A full page of advertising in a national magazine like *Glamour* could set you back tens of thousands of dollars, while a mention on the "Truth in Beauty" editorial page in that same magazine written by an editor—considered an endorsement of the highest degree—will cost you no more than your press release expenses (anywhere from a few hundred to several thousand dollars).

## A Business of Relationships

PR is a function of information (you provide) and relationships (that you develop with the press). Much of this book focuses on how to present your information to attract attention, but the relationship aspect of PR is almost just as important.

If you work with a PR agent, he or she will have established relationships with the press that are used on your behalf to make introductions and open doors. An agent can invite an editor to lunch and get a "yes" right away if the relationship is strong; it might take you 6 months to work up to that point if you start with cold calls. An established PR agent can also get the ear of an editor on your behalf and call extra attention to your press kit.

If you opt to run your program yourself, it's important to begin cultivating your own relationships immediately. Write letters, make phone calls, invite editors to your salon, attend events in your town that the editors frequent, and ask clients or mutual acquaintances to make introductions. Be consistent, but don't be pushy and don't be a pest. Be brief and get to your point immediately. If you can't get through to the head editor, ask for an assistant or associate. These people are often more open to taking calls, trying new sources, and cultivating their own relationships (Fig. 1-1).

## Consistency Counts

One of the most important points for you to learn is that to be effective, PR must be an ongoing process. When you put together your plan, you need to schedule some type of activity at least every 2 months. Every 4 to 6 weeks is even better. That's not to say you have to do photo shoots that frequently. Working with a publicist, you can develop various ideas so that your message is broad based. Look at PR as a process of creating awareness over a long period of time, rather than as a one-time event that's going to bring overnight success.

## What Public Relations Can't Do for You

Although PR can be an important and powerful communications tool for your business, it is not magic. PR can only project an image as good as, but never better than, the salon, hairdresser, or product itself. PR will enhance and strengthen what you are, but it won't make you what you are not. And it won't make you anything overnight. The bottom line: when telling your story, stick to the truth.

**Fig. 1-1** *Consistency is the key to successful PR. Stylists who send out photo shoots regularly are often featured in style pages, like American Salon's monthly "Hot Shots" section.*

# Some Terms to Know

## Model and Photographer Releases

These are legal documents that are signed by the models and photographers hired for your photo shoot that give you the right to use their work or their image in your press. These releases must accompany all press kits you send out that contain photos, for example, your trend predictions.

## The Exclusive

Once you start sending out news and features, offering your story as an "exclusive" to one publication at a time can help garner extra press (Fig. 1-2). An exclusive is a story that you offer to one publication only and do not release anywhere else until the first publication refuses to use it. You can release it to a trade, a consumer, and a local publication simultaneously. Just don't release it to competitors simultaneously. Editors love exclusives, so always let them know the story is exclusive. If it's good and appropriate for the publication, the exclusivity will increase your chances of getting your message published.

### Pro's Tip

*Give the editor a reasonable deadline for using the exclusive. Then, if the first publication chooses not to use your story or photo, you will still have time to offer it to another publication.*

One important point about exclusives: Never tell an editor a story is exclusive if it's not. If you accidentally give more than one editor the same story as an exclusive, call them as quickly as possible and admit the mistake and apologize. Although it might lose the story for you right then, the editor will remember you for your integrity and honesty and might try to work you in again.

**Fig. 1-2** *Offering an editor exclusive rights to a photo is a good way to form a relationship. SOPHISTICATE'S HAIRSTYLE GUIDE often features exclusive work from stylists in its "Tour of Beauty" section and Style Directory.*

## The Pitch

A pitch is how you present your idea. It means to suggest an idea to the editor in a format that is usable to that publication. When your publicist says, "I'm going to pitch the editor on your new hair color technique," it means he or she will try to present your technique in an interesting way that would fit right into that magazine's editorial.

## Art

When an editor asks if you have any "art," he or she is usually referring to slides, photos, or illustrations that can be used to accompany and support your story.

## Freelance Writers

Freelance writers work independently from staff writers and are hired by a variety of publications—often on a regular basis—to write articles and research various topics. A freelance writer can be a valuable source for you because he or she can present your message to a variety of publications.

## Press List

This is your current list of editors, writers, and television or radio segment producers, plus all contact information and notes on interactions you've had with them. In the back of this book, we have included lists to get you started. Just remember to verify the names, addresses and phone numbers before you use the lists.

### Pro's Tip

When calling to update the press list, get your information from the receptionist, secretary, or assistant. Try not to bother the primary contact.

## The Masthead

This is the column or page in the magazine that lists the names of the editorial staff and often freelance writers, as well. It also gives the address of the publication and sometimes the telephone and fax numbers. It's usually after the table of contents, but sometimes you'll find it in the back of the magazine. Use it to double check your press list every month and to see if it lists any freelance writers to add.

## Segment

A segment is an all-inclusive part of a TV or radio program. It usually ranges anywhere from 30 seconds to 10 minutes, depending on the format of the program, who the other guests are that day, and if the show has any themes. You will usually pitch the producer or segment producer (the TV/radio equivalent of an editor) on an idea for an entire segment.

## Lead Time

This is how far in advance an editor works. For example, the lead time for most trade magazines is about 3 months. The lead time for local newspapers can be 2 weeks to 2 months. Many consumer fashion and beauty magazines have a lead time of about 3 to 4 months; consumer service magazines have a longer lead time yet of up to 6 months.

## Clippings or Press Book

Clippings are the editorial mentions you get as a result of your PR efforts. If you have an agency or a publicist, that person will usually keep your clippings in a press book for you. You'll also want to have your own press book and several copies for your reception area and staff room.

## Consumer Magazines

Just as the name implies, these are publications that are read by the general public. They're what people buy on newsstands and through subscriptions. Most that you will target are divided into two categories: fashion/beauty and service. Fashion/beauty magazines are more trend related and tied to fashion, whereas service publications are more tip related.

Information you target to them must be consumer friendly. For example, you probably wouldn't describe exactly how you do an entire highlighting technique for a consumer publication unless you are telling the consumer how to do it herself at home or unless the consumer has a particular reason to know. Rather, you would discuss highlighting as a trend, talk about what types of faces are right for highlighting, what's required for upkeep, and the benefits of your particular technique to the consumer, that is, to get her out of the salon faster, give her a way to try hair color without fear of commitment, look more natural, look more fashionable, etc.

**Fig. 1-3** *Trade magazines run the gamut from full-service to specific topics, such as skin care or nails only.*

## Trade Magazines

These publications are read by other beauty professionals, so you'll want to give them information that's helpful from another hairdresser's or salon owner's viewpoint. Here, you would describe the exact details of a highlighting technique if that's your story, followed by how this technique can help salons sell more services, get more clients to try hair color, etc. The larger trade magazines are full service, meaning they cover primarily hair, but also include all other aspects of the salon business, including skin care, makeup, nail care, day spa, tanning, retailing, etc. (Fig. 1-3). Some are targeted specifically to the salon owner, whereas others try to include something for all professionals who work in salons from salon owners to receptionists to hair colorists to nail technicians.

### Pro's Tip

*The majority of press releases and ideas used by trade magazines are about hair-related topics. Therefore, if it applies to your business, it pays to cover these other topics, in depth. Trade magazines can also be specific to a smaller segment of the business, such as nails, skin care, or tanning. At least three publications focus specifically on each of these topics in the salon industry. These magazines are read by salon owners and professionals within their specialty.*

# Dos and Don'ts of Successful PR

No matter what media you want to target—local newspapers, radio and television, national consumer, or trade—some basic guidelines will increase your chances of having your message read, respected, and understood. Keep these pointers in mind at all times and you'll get a leg up on your competition.

DO   have a story to tell before you contact an editor, writer, or radio or television producer.

DO   read several issues of a publication or view a program before contacting them. Be familiar with what they are looking for.

DO   make sure you subscribe to any publications you are targeting. Many times, during conversations, an editor might ask you if you have seen a particular piece to give you direction or guidelines for your story. Imagine how embarrassing it would be to have to tell the editor that you don't get the publication! In addition, once you start your own program, you'll want to read the magazines closely to look for your publicity. Although some magazines will notify you, many won't, so you'll have to watch for your own material.

DO   keep in mind that the key to getting published is to send ideas, photos, and news that are of interest to a particular magazine's readers. No matter how interesting you think something is, it won't be accepted if it's not appropriate for a particular magazine. If you send something that's totally inappropriate, you risk alienating the magazine's editors by showing that you are not attuned to their needs.

DO   work to develop a relationship with an editor.

DO   make it known that you are always available to provide information and answer questions for an editor without expecting anything in return. If you're "on call" as an information center, the editor will often give you special consideration at other times when your information is appropriate for publication.

DO   call a publication. Most editors like to hear from you as long as you have something to offer that's of interest to their readers.

DO   get to your point immediately when you call an editor.

- **DO** put your best face forward when contacting an editor. Be friendly, courteous, succinct, and helpful.
- **DO** pitch new and fresh ideas when possible.
- **DO** pitch ideas that you are capable of executing. For example, if you want to talk about a new haircut that goes with the newest fashion trend, make sure you know you can execute the cut just in case the editor wants you to do it.
- **DO** be aware of how far in advance editors are working. For example, if you're pitching an idea for holiday hair to consumer editors, you'll want to approach them mid-summer. A good lead time for local media is 2 weeks to 2 months, depending if your information is more "news" or "feature" oriented. For trade editors, lead time is about 3 months. For national consumer editors, allow about 4 to 6 months.
- **DO** target publications that are in line with your salon's image.
- **DO** type all press releases or credit information.
- **DO** type all envelopes or mailing labels.
- **DO** make it easy for writers and editors to contact you. Include complete contact information, including your name, address, and telephone and fax numbers. Include an "after hours" telephone number—or "before hours" number if you're on the West Coast and want to work with East Coast editors—and note if it's applicable for Mondays, if your salon is closed that day.
- **DO** be readily accessible to editors. If an editor calls you when he or she is on deadline and needs a short quote right away, calling back 3 days later won't endear you to that person.
- **DO** educate your receptionist how to respond to media who call. Make sure whoever answers your phone knows how to convey to an editor that he or she recognizes the call is extremely important. Have the receptionist give an exact time you'll call back—that day— if at all possible.
- **DO** get a fax machine at home or your salon. Models are now available for less than $200 and they can be hooked up to your regular telephone line. Many times editors will ask you to fax information or they will want to fax information to you. In addition, if you're working with a publicist or writer, you'll need a way to send information back and forth easily.

- DO  keep copies of everything you send and to whom you sent it.
- DO  offer exclusives.
- DO  promote your press within your salon by placing framed pieces on your walls.
- DO  share the limelight with your staff whenever possible.
- DO  use your press to convey to your staff why working at your salon is great.
- DO  send color slides or transparencies. Good quality black and white photos are also usually acceptable. When in doubt, ask!
- DO  include all credit information spelled correctly when sending slides and photos.
- DO  include model and photographer releases for all photos.
- DO  label every slide and photo with your name and phone number. To label a photo, write or type on a label before attaching it to the print. Otherwise, you risk damaging your photo.
- DO  protect slides by sending them in individual slide protectors. Then, place them in a media folder or surround them with extra cardboard.
- DO  send a high-quality copy of your slides. Here's where it's smart to pay for quality. The editor will call you if the original is needed.

- DON'T  allow your staff to interrupt you while you're doing an interview with an editor.
- DON'T  ever try to push an idea on an editor who says, "I'm not interested."
- DON'T  be a pest. Sometimes even the best-meaning salon owners get so enthusiastic about an idea that they really annoy an editor. Keep your mission in perspective.
- DON'T  get discouraged if an editor says, "no." Gently try to find out why. It might just be that the timing is wrong, rather than that your idea is bad.
- DON'T  hesitate to share the glory of PR with your staff.

**DON'T** send poems, short stories, or other "creative writing" pieces to trade or consumer magazines unless you're certain this is something they are seeking.

**DON'T** send color print photos. They do not reproduce well in magazines.

**DON'T** send original slides unless they are requested.

**DON'T** send loose slides. They will be damaged in transit and make your presentation look unprofessional.

**DON'T** write directly on the back of a photo when labeling it. This damages the print and makes it difficult to reproduce in the magazine.

# Chapter 2

# Honing Your Salon's Image

*F*irst impressions are everything. You never get a second chance to make a first impression. That old axiom is just as true whether you're referring to the first moment a client walks through your doors or the first telephone call made to an editor or the first press kit sent to your local newspaper's fashion writer. You need to make sure that your image is represented accurately to present your message properly.

When you send a press kit to a trade magazine editor, you might be competing for attention with at least 35 other press kits she receives each day. When you're targeting the consumer press, the competition can get into the hundreds—per day! You have to have something special that makes you stand out. The goal is to create a professional image. A professional image not only gives you the credibility you need when dealing with the press on both the local and national levels, but also expresses a self-confidence and belief in your products, service, staff, and salon.

## What Is Your Salon's Image?

That brings us to the big question: "What is the image I want to project?" That is followed by, "Can I back up that image?" and "Can I pull it off in my presentation?" (Fig. 2-1)

First, define your target client. Write down a detailed description of the type of client you wish to have in your salon. This is the person you will speak to in your PR, your advertising, and your promotional efforts.

**20** Honing Your Salon's Image

**Fig. 2-1** *Noëlle Spa for Beauty & Wellness in Stamford, CT, has a calm, serene, spiritual image, as reflected in its design and decor. Those same qualities are present in the salon's press materials. Photo by Janet Durrans.*

Establish your salon's unique selling position. On the same piece of paper, write down what makes you and your salon different in the minds of your clients and your staff. Why do clients patronize your salon instead of the one down the street?

Assessing your strengths honestly is important before proceeding with a publicity program. For example, if you are quoted in an article in *Glamour* touting the variety of hair color services available today, you might receive literally hundreds of telephone inquiries from potential clients. You must make sure you can handle them—at your reception desk and in the chemical services department.

## Success Story: The Hairmeister of Las Olas is on Call to the Press 24 Hours a Day!

Salon owner Jesse Briggs spends at least an hour every day making a name for himself in the local, national, and international media—in addition to the 8 hours he spends behind the chair.

Jesse's PR efforts begin at Yellow Strawberry Global Salon in Fort Lauderdale, which he owns with his wife Flo. Although most salons fill their walls with photos of the latest hair styles, Jesse puts his own personality into the decor, and people come from miles around to see his collection. The walls are filled with memorabilia from entertainment to politics collected for more than 30 years. His pride and joy is his collection of locks of hair of famous people, including Elvis, John Lennon, John F. Kennedy, George Washington, John Brown, Napoleon, and Robert E. Lee. The hair has earned him recognition from Los Angeles radio to CNN.

Other items on display include a suit, shirt, and tie owned by Jack Ruby, a collection of Robert Frost's original poems, Irving Berlin's picture and autograph, Kate Smith's/Irving Berlin's signatures on the music of "God Bless America," and much more.

Because Jesse loves to share information on every topic, he has been named the Hairmeister of Las Olas by his local media. He'll comment on everything, and reporters rely on him giving an amusing twist to any topic. In fact, the day that Jimmy Johnson was named coach of the Miami Dolphins football team, a Miami Herald reporter was on the phone immediately to ask Jesse to comment on Jimmy's hair, as well as that of all the area coaches.

Following are just a few of the additional PR strategies that Jesse has used recently to keep his name in the limelight. These ideas required an investment of time more than money, combined with Jesse's creative thinking and a mind that's always open to opportunity. Jesse Briggs ...

- ... was one of the first salon owners to set up his own page on the World Wide Web, where consumers can find his recent styles and beauty tips and order products.
- ... joined forces with a local trainer to cohost the cable TV show, "Beauty & Fitness USA."
- ... sponsored Mary Ellen Clark, an Olympic veteran and the premier US woman's diver, and Kelly Moore, ranked as the top US windsurfer. As a result, the Yellow Strawberry logo has appeared on Mary Ellen's bathing suit, Kelly's wind surfing sail in competitions around the world and in profiles of the athletes featured on TV and in "USA Today."
- ... sponsored an area man for his appearance on the TV show "Star Search." In return for the funding, the man wore a Yellow Strawberry T-shirt during his nationwide appearance.
- ... with Flo, annually hosts the Caribbean Hair Affair, a 3-day cruise featuring hair color education.
- ... volunteered time and money to head up the industry's Hair Cares Angels program.
- ... does semiannual photo shoots with Yellow Strawberry staff

> members from around the world. These shoots give him enough material for an entire year. Pick up any consumer hair styling publication and you'll find Yellow Strawberry's work.

## "What's My Image?" Worksheet

Before you can expect anyone to write about you, you first must clearly define who you are and what kind of image you want to promote. To help you assess your image completely, we've created the "What's My Image?" worksheet. This is designed with a twofold purpose. First, it helps you get your thoughts in order. You can also make copies and pass them out to your key staff members. See how their answers compare with yours. If there's a wide difference, then something's wrong and you have to both check your image and your staff's understanding of that image. What a great topic for a staff meeting!

Once you have agreement, fill out a master worksheet. You can use this for reference each time you do a PR, promotional, merchandising, or advertising activity to make sure your message is in sync with your desired image.

This will also be an invaluable tool to give to your PR agent if you decide to work with one or to the freelance writer who will help prepare your press releases and collateral materials. The answers to these questions will give the agent a strong starting point and a thorough understanding of how you see your business and what you want to achieve with your PR program. The PR professional can then incorporate your concepts into the overall plan for you and your salon.

PR magnifies what you already have so be sure all of the components that contribute to your overall image are aligned. Those components that make up your salon's image include your salon's physical appearance, your staff, the level of service you offer, your pricing, your logo, and all of your collateral materials, including your letterhead and service menu.

Here are a few questions to ask yourself:

**Who makes up the majority of your clientele?** (Be as descriptive as possible and note if there are several major groups. A few examples are young, old, professional, student, conservative, society, hip, etc.)

_____

_____

_____

**Does your salon reflect this client profile in atmosphere, decor, and service?** (For example, if you service your area's professional working women or executives: Are you efficient? Can clients be in and out of your salon quickly? Do you offer them plenty of quick yet de-stressing services? Do clients have access to telephones for business calls? Does your staff dress in a way that says they are fashionably professional? Is your decor more functional and tasteful rather than elegant and fancy? If you service men, is there an area where they feel comfortable?)

_____

_____

_____

**Do you want to attract more clients like you have today? Or do you want to attract a very different group?** (Be as descriptive as possible about the new group you want to attract. Who are they? What is their age range? Income? What types of publications might they read?)

_____

_____

_____

**What are your most popular services currently? Why?**

_____

_____

_____

**What are the services that make you the most money?**

_____

_____

_____

**How well do your logo, letterhead, business cards, and salon price menu reflect your image?** (For example, if you want to attract the "society" ladies in your area, your business cards and logo should not look too trendy or high tech, but rather be more classic and elegant.)

_____

_____

_____

**I never want to be seen as....**

 

**If consumers in your community could know you for only one thing, what would you want it to be?**

 

**If other salon owners and hairdressers could know you for only one thing, what would it be?**

 

**Name at least one important way that your salon is different from your competition.**

 

**What was the last big "buzz" in your salon about?**

 

**What charities is your salon currently involved with?** (Do you tend to favor particular causes that support your image and give you the opportunity to meet more of your desired clientele? Or do you tend to consider any request for help that comes up?)

What magazines do you have in your reception area? Are they in tune with your image and are they the publications you want to be seen in?

_____

_____

_____

Once you've answered these questions, you should be able to clearly define your salon's image. If any one of these elements doesn't fit or contribute in a positive way to that profile, then seriously consider adjusting it to avoid presenting a confusing or unfocused image to the public. Remember, consistency is everything.

# Collateral Materials

Once your image is clear to you, you will need to make sure it comes across clearly to the public in all of your collateral materials that will be used within your PR program.

Often, if you're starting from scratch, it can be expensive to have all of the pieces designed and printed at once. However, because they will be similar in design and color, it will likely cost less to have them all designed and printed at once than if you have it done piecemeal. Try to negotiate a discount with the graphic designer and printer because they will be getting a great deal of business from you at once.

If you don't already work with a designer and printer, comparison shop once you know exactly what pieces you will need. Compare both quality and price. This is not the time to skimp on quality.

There are several ways to find a designer who has experience in putting together a "corporate identity," which is what the designer will be creating for you. Ask your clients who their companies work with. Study other businesses in your area and find those with a sense of style. (That doesn't mean you will copy their design. It just means that you like how all of the elements work together.) Then, ask who they use. You can also put up notices at local colleges or art schools. If you have a printer, you can ask your contacts there to recommend someone because they know who they like to work with.

*A word of warning:* Many people call themselves designers simply because they have a Macintosh computer, but having a computer does not give them graphic talent and a sense of style. No matter how you find a designer, make sure you ask to see several examples of work he or she has done, especially in the area of "corporate identity" packages.

**Fig. 2-2** *Salon owner David Kinigson had a unique logo designed for Salon Dada that visually communicates its unique point of view.*

You can go about finding a printer in a similar manner. Or, find your designer first and then ask the designer who he or she likes to work with.

If you are going to work with a PR agency, this company most likely has resources in both areas. Don't be afraid to ask for their advice. Also, when putting together the pieces discussed below, bring in the agency for advice when making your final selections. After all, you are paying them for their strong sense of what editors like to see. Use their knowledge to your advantage! (We'll discuss how and when to work with an agency in the next chapter.)

Here are the pieces that must be in place before you begin:

## *Logo*

You may already have a logo (Fig. 2-2). Now is the time to look at it in light of your focused image and make sure that it truly supports the message you want to convey. You also want to make sure that it appeals to your target clientele. For example, if the majority of your clients are professional women and that's the type of clientele you want to continue to attract, you don't want a logo that's super hip. Likewise, if you primarily service the "Club Kids," who are beyond the cutting edge of fashion, you don't want a logo that's too conservative. Ask your graphic designer to evaluate your current logo in light of your image. See if he or she can suggest some modifications to make it appealing to your target audience or if something else would be more appropriate.

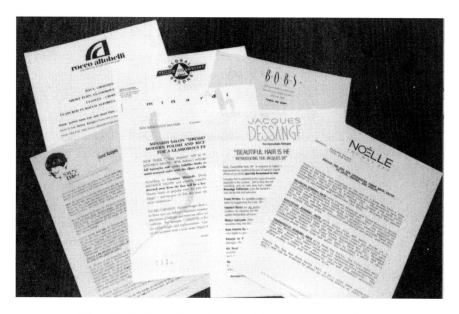

**Fig. 2-3** *Samples of salons' logos and letterhead*

If you don't already have a logo, talk to your graphic designer about creating one. Give the designer a copy of your image worksheet so he or she can see exactly what you are thinking about your salon's image. Then, have a face-to-face discussion to clarify any other points. Give the designer specific words to define your salon's image to use as a guideline. Examples would include hip, elegant, fashionable, quick service, cutting edge, and so forth.

## Stationery or Letterhead

This is the 8 1/2" x 11" paper that has your logo, as well as your salon's name, address, telephone number, and fax number (if you have one) (Fig. 2-3). As with the logo, if you already have stationery, this is a good time to give it a second look.

In addition to your logo design, you'll have decisions to make, such as will you use only black ink or can you afford two colors? What type of paper should you select? Remember, even your paper selection must fit with your image. For example, do you want a paper that's upscale and elegant? Or an earthy recycled stock? Do you want a color, a speckletone, or an off white? These are all decisions to discuss with your designer and printer. They are best positioned to advise you about feasibility and cost both of setup and reprinting. When in doubt, from the standpoints of both cost and presentation, it's usually better to tone it down a notch than to overdo it. It can also save you some money.

## Pro's Tip

*If you're working with a PR agency, you might opt to have your press releases sent out on the agency's letterhead. This is a good idea if you're paying for a big name agency and want to use that agency's reputation to create immediate name recognition for you. On the other hand, if you're working with a lesser known agency who still has some contacts, the cover letters might be sent on the agency's letterhead while the press releases are sent on your salon's letterhead. Either way, you'll want to include this in your budget because you will be charged for use of the agency's letterhead. Regardless, you will still need your own business cards, salon menu, and other materials that must come from your business.*

### Press Folders

Press folders are usually 9" x 12" or 10" x 13" in size (Fig. 2-4). They have pockets and a place for your business card. They will be used to hold your press releases, photographs, and other collateral material. If budget isn't a major obstacle, you can have customized folders designed to match your letterhead. The next best solution is to go to a stationery

**Fig. 2-4** *Press folders hold your PR material.*

store or an office supply store. Find a folder that is consistent with your letterhead and logo and buy the quantity you need. Printed labels with the logo and name of your salon, affixed onto the front of the folders, is a less expensive way to get a customized look.

As with letterhead, if you are working with a PR agency, you might end up using their folders. Once again, you will be charged for their use.

## *Additional Printed Material*

This includes your salon menu, business cards, flyers, and any other items that you need to complete your image package (Fig. 2-5). You will want to be sure that all of your printed material carries your logo and creates a consistent image or corporate identity.

### Pro's Tip

*Many editors like to write notes on the front of press folders. You'll make their job easier if you select light colored folders in a matte—not shiny—finish that can be written on easily with ball-point pen.*

**Fig. 2-5** *A salon's printed material should coordinate.*

## Pro's Tip

Often, editors will ask to see your menu, so it must be in line with the rest of your materials.

# A Word About Photo Shoots

One of the strongest ways you will convey your image is through your photo shoots. In fact, at least twice a year, you will probably rely on your photos to create your primary message, which is then supported by your letterhead and corporate identity materials.

Although you have probably done photo shoots before, it's time to re-examine them in light of your newly fine-tuned image. If what you're saying through your photos isn't consistent with what you want to say as a business, then it's time to make some changes.

Milady's book, *Picture This...Your Image in Print*, is an excellent source for both the photo shoot expert and novice. It gives you step-by-step guidelines for expressing your image through photos, as well as selecting the right people, from photographer through makeup artist, to make sure your vision comes to print just the way you want it. If you're serious about getting your image together, it will be a useful resource guide for you.

Victoria Wurdinger, author of *Picture This...Your Image in Print*, offers this advice:

> If you want to set up a photo shoot and use the shots for public relations purposes, start by observing what types of photos your targeted magazines are running. If you want to get your work published in the local newspaper, you'll want to create an image that's uniquely yours, but that also makes sense with current spring/summer or fall/winter fashion trends.
>
> Start learning about shoots the same way you learned about running your salon. Observe others, network with people who have done shoots and start asking for advice. Ask a local photographer or makeup artist or even another business owner if you can sit in on a shoot and observe. Perhaps the bridal shop next door to you is creating an advertisement. Offer to do the hair or simply ask if you can be on set if you don't interfere. Then, make notes of everything you see.
>
> Sometimes, other salons don't mind sharing information if you aren't a direct competitor. If a salon across town lets you sit in on a shoot, tell the owner you'd like to explore the possibilities of an eventual joint session, which will save you both money. Once you're confident, you can pool resources to hire top talent and each shoot two models or pay a full day rate for specialists and split the day in half between salons.
>
> A photo shoot must have a specific purpose and be well-planned in advance. The more places you can use the photos, the better! It should involve your salon team, but only those who are good at styling hair for shoots, doing makeup for photography and so forth. Don't use a shoot to reward longevity or loyalty; make it clear that in this case, the most appropriate person is the one who will be chosen.
>
> Makeup for photography is very different than real-life makeup; "editorial" hair is not necessarily good salon hair. Ask photographers and other professionals to describe the difference to you. Look at photographers', stylists' and makeup artists' portfolios and you'll begin to see why experience behind a camera matters so much.
>
> If you don't think you or your staff is experienced enough to create a shoot to be used in a press kit, hire outside talent. Most salons use professional makeup artists, whom they book through an agency. You can also use a professional session stylist. Just make certain that you aren't totally misrepresenting what your salon can do. If local papers run a feature on your new perms, it's fine if an editorial stylist can make them look their very best in print, but make certain you can make them look good in real life, in the salon.

Finally, keep in mind who uses photo shoots from salons and who doesn't. Local newspapers often send their own photographer, but will use shots usually in black and white. Salon industry trade magazines are always looking for finished styles, step-by-step technicals, new product shots, trend statements and even photos of salon interiors. Major consumer magazines will not print photos sent to them from salons. They always shoot their own material. They will review your photos to get an idea of your expertise.

## Model and Photographer Releases

These are legal documents that are signed by the models and photographers hired for your photo shoot that give you the right to use their work or their image in your press. These releases must accompany all slides and photos that you send out. (You'll find samples in this chapter.) These releases are important and most publications will not even consider publishing your photos without them.

### Pro's Tip

*Have the photographer and model releases signed on set before your photo shoot begins. First, you want people to know exactly what they are agreeing to before they are photographed. And it's much more difficult to try and track people down later.*

## Model Release Form

In consideration of the sum of $1.00 and other valuable considerations mutually agreed on and received by me, I, the undersigned, hereby authorize _____, and such other persons, firms or corporations as may be acting on behalf of _____, to photograph me and use any and all such photographs, without limitation as to time or method of reproduction or exhibit, for all purposes, including but not limited to advertising of all kinds in all media of advertising, promotions, or merchandising, initiated, sponsored or approved by _____ and his assigns.

Date: _____

Name: _____

Age: _____

Street Address: _____

City, State: _____

Zip Code: _____

Telephone: _____

Signed: _____

### Guardian's Consent

I, the undersigned, as parent or legal guardian of the above-named minor, do hereby agree and give my consent to said minor entering into the above agreement and do guarantee thereof by said minor.

Date: _____

Name: _____

Street Address: _____

City, State: _____

Zip Code: _____

Telephone: _____

Signed: _____

## Photographer's Release Form

I, _____ (photographer's name), resident at _____ (studio location), hereby irrevocably consent and agree that _____ (the hair stylist and makeup artist names) may authorize or represent, and other in private with _____ (the hair stylist and makeup artist names), may use, employ, reproduce, and distribute, in advertising literature, trade publications, trade journals, and/or instructional or promotional material, in any form deemed appropriate by _____ (hair stylist and makeup artist names), my name and/or any or all of the pictures/photographs I have submitted or have taken at the authorization and expense of the same, and I hereby waive any and all rights to inspect or approve such material or use. I have made no agreement, limitation or commitment to any other or others that prohibit me from entering this agreement.

Name: _____

Date: _____

Witness: _____

> **Note**
>
> *These release forms are samples only. Contact your attorney for approval before using them. In addition, keep in mind that most professional models will not give away advertising and promotional rights without a substantial additional fee.*

Now that you have decided what your salon's image is and what message you want to present to the world, it's time to find out how to introduce yourself to the press.

# Your First Press Mailing: The Backgrounder

Your first impression will come through an initial press kit called a "backgrounder." This is a complete press kit that tells everything possible about you, your salon, and your staff that's relevant to your PR program. It conveys what you are all about and gives the media the facts and figures that tell them why they and their audience will find you interesting and unique. It may contain some or all of the following elements packaged in your media folder.

- ◆ Your Salon's Story: This is a brief press release describing your salon and what makes it unique, including your salon's image, staff, and target clientele; its unique niche; its positioning in your town; any special activities you participate in (your "What's My Image" worksheet will guide you about what to include). (1 to 2 pages maximum)
- ◆ Fact Sheet: This is a list of important facts about your salon presented in a straightforward list. (1 page only)
- ◆ Bio: Include biographies of you and any of your top staff members who will be featured in your PR. Talk about important experiences

and areas of expertise. (Try to keep it to 1 page—see sample in press releases section.)
- ◆ Photos/Slides: Include sample photos from you or your artistic team to give the editor an idea of your capabilities and your vision of beauty and fashion. (Note: Send actual publishable work to consumer hairstylers, local press, and trade magazines. Send clear-reading color photocopies of your styles to the general consumer media because they just want to see your ideas. They won't actually publish the work. Do whatever is most cost effective.) Always send model and photographer releases with any slides or photos (see example, this chapter).
- ◆ Menu: Include your salon's menu of services and prices.
- ◆ Business Card: The media can pull it out and add it to their source files.
- ◆ Press Clippings: Include any particularly relevant or recent press clippings that present you as you want to be seen.

### Pro's Tip

*Your backgrounder is an evolving kit that should be updated to reflect any changes in your salon, including additional services, new accomplishments, etc. However, it's a good idea to have five complete kits on hand at any time. Often, an editor on deadline will call and want you to overnight or even fax information. You do it or lose the opportunity. You won't have time to have slides duplicated or to write an extra press release.*

- Newsletter: If you do a salon newsletter, include your most recent copy.
- Cover Letter: Although this won't go within the folder, include it in the envelope to briefly tell the editor who you are and what you're sending. Make sure each cover letter is personalized with the editor's name and address. "Dear Editor" is not acceptable.

Keep in mind: Once you make your point, less is more. Don't overwhelm the editor by sending materials that don't support your primary story.

After the backgrounder is out, you'll continue to give the press something to write about by sending out press releases. A grand opening, a new service, your fall trend forecast, and hints for home care are all good stories—if presented in the manner that's most useful to the media you are sending it to. The same story might work for both your community newspaper and *Vogue*. The difference will be in how it's positioned to each. (We'll discuss that difference in detail in the following chapters.)

## Press Releases

Your press releases will fall primarily into three categories:
1. The Trend Release: This is your prediction for hair, beauty, and fashion trends for the upcoming season. It's the written statement that accompanies your photos to tell the editors why you created the hairstyles you did in this particular photo shoot.
2. The News Release: News is something that's current. It is usually time specific. For example, if you're holding a cut-a-thon on April 12, that is news. Because of time constraints faced by most magazines—by the time they can publish it, it's no longer news—your news will be targeted primarily to your local press, with some to the trade press.
3. The Feature Release: You can also think of this as a "tip" release or a "story." Usually, time is less crucial to a feature. But, it might still fit in better with one season than another. For example, an idea about how to wear the newest updos is a feature. Make it seasonal by targeting it to weddings, proms, or the winter holidays. The same goes for makeovers. The media usually love them—just be aware of who prefers subtle changes to dramatic redos. Then, makeovers can be targeted to any season from Mother's Day to Back to School.

Following is a sample bio that is included in Yellow Strawberry Global Salon's backgrounder and is sent out when the media request information.

**FOR IMMEDIATE RELEASE**
Contact:  Jesse Briggs/Yellow Strawberry Global Salon
(555) 222-1234

## JESSE AND FLO BRIGGS
## THE MOST COLORFUL COUPLE IN THE BEAUTY INDUSTRY

The most colorful couple in the beauty industry, Jesse and Flo Briggs are two of the world's most renowned hair colorists, salon owners, educators, entrepreneurs, and innovators.

Unlike any other salon owners, Jesse and Flo—husband and wife for 35 years—head a growing beauty enterprise that is rapidly spanning the globe. In addition to the flagship salon in Fort Lauderdale, Jesse and Flo own or are partners in Yellow Strawberry Global Salons located in Miami Beach, FL; Sarasota, FL; Key West, FL; Huntington Station, NY; Knoxville, TN; Puerto Rico; Paris, France; Florence, Italy; Montreal, Canada; and Tripoli, Lebanon, as well as their Yellow Strawberry Global Salon in the Fashion Mall in Plantation, FL. That salon is run by their daughter, Denise Briggs, who is president of the company. Plans are underway to open more Yellow Strawberry Salons around the world this year. In total, Yellow Strawberry Global Salon employs 900 stylists and colorists, plus other beauty professionals.

In 1993, Jesse and Flo were honored with "The Oner" (The Absolute) Award for their contributions to the global salon industry. This prestigious international honor is presented yearly to "the expert in creativity and professional ability in the world of hair styles," by *La Griffe* magazine, based in Milan, Italy, and its Chief Editor, Attilio Fregoli.

As International Artistic Directors for BES Regal Hair Color, the Briggs head up that company's education programs in the United States. Respected as master educators around the world, they recently opened the Briggs & Briggs Hair Color Institute to advance the technique, artistry, and business of hair coloring. In 1996, they will host their 12th Annual Caribbean Hair Affair (hair color education aboard a cruise ship), and they are cofounders of the prestigious HairColor USA.

The Briggs' beautiful hairstyles are published worldwide, and they are quoted around the world for their innovation in developing new techniques, such as "Ribbon Coloring," a technique for creating beautiful, natural-looking highlights without foil. They also take pride in bringing hair color out of the back room to the front of the salon.

"At Yellow Strawberry, color chemists mix clients' formulas right in front of them to take the mystery out of hair color and bring the artistry of the service and technique center stage in the salon," say Jesse and Flo.

Always focusing on total beauty, Jesse and Flo were recently named US spokespersons for DIBI USA's Dibitron Face System. Their salon has been named as a DIBI Model Centre, where Dibitron Face techniques will be showcased and taught.

Jesse and Flo have been the recipients of more than 500 plaques of appreciation and European hair colorist awards, and they have also appeared on television and radio interviews worldwide. Their work has been featured in top publications, such as *Vogue, Glamour, Harper's Bazaar, Elle, Mademoiselle, Seventeen*, and *Ladies Home Journal*.

These extraordinary individuals will be remembered as two leading forces of the 20th century, not only for their accomplishments but also for their character. Flo is a multitalented woman whose focal interests include art and culture. Jesse is a true philanthropist, who by using his keen business intuition, his willingness to overcome challenges and his charismatic personality enjoys life by helping others succeed. As one example, he is director of the HairCares Angels program, a division of the HairCares/BeautyCares charity that helps beauty professionals who have AIDS.

Yellow Strawberry is also spanning the globe via the Internet. The address is http://www.bizflorida.com/bizflorida, where you'll find a listing of locations, products and beauty tips.

# Sample News Release

**FOR IMMEDIATE RELEASE**

Contact: Mario Tricoci/Mario Tricoci Hair Salons and Day Spas    (555) 222-1234

**TWO NEW WONDERFUL SERVICES—THE PUMPKIN PEEL WRAP AND OXYGEN TREATMENT FACIAL—HELP REVIVE WINTER WEARY SKIN**

Chicago—DATE—Just in time to help your face and body skin bounce back from the effects of chilly winter winds, drying indoor heat, heavy winter fabrics, and a less active lifestyle, Mario Tricoci Hair Salons and Day Spas have added two new spa services that revive and rejuvenate dry, dull, and sluggish skin. The 26 massage therapists and 26 estheticians have undergone hundreds of hours of training under the direction of Spa Director Fabienne Guichon to make these wonderful services available just in time to help you look and feel your best for more revealing spring fashions.

1. PUMPKIN PEEL WRAP Imagine the wonderful fragrance of pumpkin pie in a body treatment that's also one of the most effective available. Our new Pumpkin Peel Wrap is packed with 10% to 15% natural enzymes—today's hottest skin care ingredients—that gently and naturally exfoliate the skin by dissolving away the dead surface cells. It leaves your skin silky smooth and is especially beneficial for people with sensitive skin that can't take the physical exfoliation of a mechanical scrub. As an added benefit, the Pumpkin Peel Wrap may help control psoriasis and eczema. Cost is $60 for a 30-minute treatment.

   Extra benefit for spring: Removes dry, dead cells accumulated during the winter, revealing softer, smoother, brighter skin.

2. OXYGEN TREATMENT FACIAL To live, we need water, oxygen, and nutrition. That's what our skin needs, too. But as we grow older, our circulation begins to slow down, which slows the flow of oxygen and nutrients to the skin cells. To keep skin looking healthy, we've developed this three-step anti-aging service to get oxygen deep into the skin to feed the cells with oxygen and nutrients.

   If you're over 40, you'll see wonderful results and firmer skin with a series of six at each change of season. If you're in your 30s, get a service once a month. Once every other month provides great maintenance if you're in your 20s. Cost is $75 for a 45-to-60-minute treatment.

   Extra benefit for spring: The Oxygen Treatment Facial is the perfect treatment to rev up skin that's become dry, dull, and sluggish from a winter of exposure to drying winds, drying indoor heat, and lack of exercise. It's also a great treatment to help improve the skin of smokers.

To maintain benefits at home, Guichon recommends Tricoci's Antiox Serum, an antioxidant cocktail rich in ascorbic acid (vitamin C) that's worn daily on the face and neck to help neutralize free radicals and minimize the aging process' effects on the skin. It helps to refirm the skin, leaving it looking healthy, youthful, and beautiful.

Mario Tricoci Hair Salons and Day Spas are located in downtown Chicago and its surrounding suburbs, including Schaumburg, Oak Brook, Bloomingdale, Arlington Heights, Naperville, Old Orchard, and Crystal Lake, IL and Kansas City, MO. Watch for the new Day Spa in Orland Park, IL, soon! For more information on services, products and spa gift certificates or the day spa near you, consumers can call 1-800-444-1234.

# Sample Trend Release

**FOR IMMEDIATE RELEASE**
Contact:   Robert LaMorte/Robert Jeffrey Hair Studios   (444) 212-1234
**EXOTIQUE
PASSIONATE INTERPRETATIONS OF CLASSIC STYLES
BY ROBERT LAMORTE**

Exotique is...

... alluring combinations of fashion and beauty that smolder with artistic passion.

... classic interpretations of design that are pushed to their outer limits.

... modern expressions of the extrinsic form and shape of hair designed on the canvas of the soul.

Robert LaMorte's Exotique Collection dares you to take a journey to the more alluring side of hairdressing. Experiment with new interpretations of classic hairdesign techniques and try something new! Whether it be an updated French Twist or a '90s grunge look that goes mainstream, LaMorte pushes classic techniques to their outer limits of passionate interpretation.

CREDITS

Hair & Concept:  Robert LaMorte for Robert Jeffrey Hair Studios, Chicago

Makeup:  Jane Doe

Photography:  John Doe

Fashion Styling:  John Smith

# Building Relationships

Before we get any further into the mechanics of putting together a PR program and publicity presentation, it's important to focus on the importance of building a relationship with members of the media.

Simply put, it's possible to present the exact information an editor needs and not have it used. Why? If an editor has two equally good packages of information, chances are he or she will use the one from the person who has developed the best relationship.

Here are some key points to developing a good professional relationship with the media:

- ◆ Present yourself and all materials professionally. Return calls quickly.
- ◆ Be pleasant and polite.
- ◆ Always get to your point quickly.
- ◆ Call back just once to be professional, but never pester the editor.

- Treat all editors or members of the media with equal respect. Today's junior assistant is tomorrow's senior beauty director. If you were a great source to her when she was a junior, she just might take you to the top with her.
- Never try and go over an editor's head to a more senior editor—or even worse—to a member of the sales staff. If you buy advertisements in the paper, your sales reps can be helpful in telling you which fashion/beauty/lifestyle reporter to speak to, but you should never ask the person to put pressure on an editor to include your work just because you advertise.
- If an editor says "no," thank him or her politely and get ready to send in your next idea. Show persistence and confidence in your work.
- Always be pleasant and polite to receptionists, assistants, and everyone else who works with the editor. People talk. Be rude to a receptionist and chances are, the editor will know before you can hang up.
- Show that you are willing to share information with the media, even if it might not mean a quote for you at the time. They'll appreciate your professionalism and remember it.
- Never, never, never talk badly about your competitors.
- Be knowledgeable and informed about both the media you are pitching and the topic you are pitching. For example, if you have a new hair color trend, the editor will appreciate knowing what came before that led up to this trend.
- Be knowledgeable about national trends and even international trends when applicable (especially if it's related to the European fashion runway shows) and how they relate to your trend. Or, know if your trend is strictly regional and perhaps even the opposite of what's happening on the runways.
- Drop names only if you truly have a strong relationship with the person whose name you dropped.
- Finally, be 100% honest and credible. Make sure you can live up to what you promise and don't ever promise more than you can deliver.

# Chapter 3

# Planning Your PR Program

## What Do You Want Out of PR?

The first step is to decide why you want PR. That might not be as easy as it sounds. You can use the following quiz as a helpful tool to decide what media are best for your message. Think about the following questions and rate them all using this scale:

A= very important, B= somewhat important, C=not as important.

1. I want my salon to be known as the best place to get a great cut in my town. _____
2. I want to be an international superstar in the hairdressing community. _____
3. I want to attract top hairdressers to work at my salon. _____
4. I want to be recognized as a business leader in my community. _____
5. I want to become a national household name. _____
6. I want other hairdressers to look at me as an innovator of new techniques. _____
7. I want to become well known enough to be tapped as an educator or platform artist by a manufacturer. _____
8. I want to do makeovers for national beauty magazines. _____
9. I want my current clients to see my salon as the place to go for all of their beauty services. _____
10. I want to appear as an educator at a major beauty show, such as the Midwest Beauty Show or the International Beauty Show. _____

11. I want to be recognized by my peers and my community as a successful business person. _____
12. I want my hairdressers to be proud of where they work, not just because of our great work, but also because our great work is recognized. _____
13. I want my hairdressers to get greater recognition for all the time they donate to charity events. _____
14. I would like to share my time-saving techniques with other hairdressers. _____
15. I've devoted 25 years to building this business and I want recognition! _____

## How to Score Your Answers

- ◆ If your priorities are statements 1, 3, 4, 9, 12, 13, and 15, then you probably want to focus your time and dollars on local PR.
- ◆ If you answered A to statements 1, 3, 5, 8, 9, 12, and 15 then national PR is important to you.
- ◆ Finally, if statements 2, 3, 6, 7, 8, 10, 11, 12, 13, 14, and 15 are the most important to you, then industry/trade PR is for you.

Chances are, there will be some overlap, because many of the questions themselves overlap from local to national to trade PR, and you will want some mix of the three. This will give you an idea of where you will focus the bulk of your time and dollars. Based on your priorities, you will decide what press to target first.

Also, you'll want to share this list with your PR agent if you decide to work with one. It will help make sure that both of you are on the same wavelength when it comes to goals. If your PR agent continuously gets you mentioned in *Glamour*, but what you truly want is greater recognition in the hairdressing community, then some changes will have to be made in the program.

One area we touched on only slightly with statement 15 is ego. Whether or not they think of it consciously, satisfying their ego is one of the primary reasons that people want PR. There's nothing wrong with that (you've worked hard and deserve recognition!), just as long as you keep it in perspective and can back up your "claims to fame." Combine it with a sound PR plan that's also designed to offer other business-sound benefits and you probably won't go wrong.

## Developing Your PR Program

Now that your salon's image is firmly implanted in your mind and on paper, you can begin to set your PR goals and develop the plan that will help you achieve them. Using communication that's consistent and credible, your goal is to create the correct positive public perception of what your salon has to offer and to convert its authoritative position on beauty and fashion trends, the latest techniques and products, and the most professional skills and services into a message that is easily usable by the media.

Although you might think of PR as press kits and interviews, it actually includes every contact you have with the public and the media from each contact with a client in your salon to a major makeover appearance on *Oprah*!

Some components of an overall PR program can include:

- Sending out press kits (referred to as publicity).
- Participating in charity events.
- Donating gift certificates from your salon as prizes for local raffles.
- Doing the hair for local celebrities or newscasters.
- Meeting personally with trade editors at national beauty shows and consumer editors in New York City to share your story.
- Even how you treat your clients on a daily basis is a form of PR. (In fact, happy clients who will talk about you and how much they love their hair to their friends, families, and professional groups can be one of your strongest methods of PR and it's definitely worth cultivating. In fact, it's probably the least expensive method of PR and yields some of the most powerful results.)

## Targeting Your Message

In creating your new PR program, you can combine several or all of the above activities, depending on your salon's strengths and interests. You can target your message in several ways:

- Locally, through newspapers, radio and television
- Nationally, through consumer magazines, such as *Glamour, Self, Family Circle,* and *Marie Claire*
- Within your industry, through trade publications, such as *Modern Salon, Salon News,* and *SalonOvations,* and trade shows

### Local

This means any media in your area and can include newspapers and local magazines, as well as radio and television programs. It also includes the types of charity events you are involved with, for example, grass roots client-based efforts versus the high society fund-raisers in your community. Which you choose to focus on depends on your image, as discussed in the previous chapter.

**Fig. 3-1** National hairstyle publications

## National

This includes targeting your press to national beauty, fashion, and women's service magazines, such as *Glamour, Vogue,* and *Woman's Day*. It also includes national hairstyle selectors, which are sold in grocery stores, drug stores, and newsstands across the country (Fig. 3-1).

It can also involve working on shoots if you are a session stylist or makeover artist. Finally, for the very famous, the very persistent, and the very lucky, this can include national television talk shows, such as *Oprah*. One of the best ways to get an "in" with these shows is to do the hair of a famous person who might be a guest on the show. This requires the greatest financial commitment.

## Industry

The industry trade magazines, such as *Modern Salon, Salon News, American Salon,* and *SalonOvations,* can help make you a star within your profession and a leader among your peers (Fig. 3-2, 3-3, 3-4). A mention in these magazines can also be turned into a local press release. For example, if your hairstyle is featured in *Modern Salon,* then you can send a press release to both your local fashion editor and your local celebrity or gossip editor announcing it with a copy of the placement.

Consistent, ongoing publicity within your industry can also lead to recognition through professional organizations, interest from manufacturers looking for creative consultants and educators, and interest

**Fig. 3-2** *Full-service trade magazines cover all topics: hair, skin, nails, salon design, business, and more.*

from national beauty shows who are always looking for new educators to spotlight. It can also lead to referrals from other salons for their clients who are traveling or moving across the country.

Mentions in the trade magazines can also lead to name recognition by the consumer editors, who often refer to the trade publications for reference, information, and ideas. They know that someone who has respect from peers within the industry is probably a credible source for their articles, too.

Don't overlook this important category. If you have already budgeted money for shoots and materials to do local PR, you can easily and inexpensively add trade simply by retargeting your press release.

Later in this book, we'll spend an entire chapter on each of these areas to tell you exactly how to target each, how to get your lists together, as well as how to maintain those lists. We'll also give you professional tips for successfully targeting each of these areas.

Your final plan will probably include some combination of those media. Although you do want to focus your message, in many cases, it doesn't make sense from an expense standpoint to limit your audience. For example, once you do a photo shoot to send to your local press and use in collateral materials, the additional cost to send it to the trade press is minimal. But, no matter what audience you choose to target, all areas must work with one another and support a consistent image and

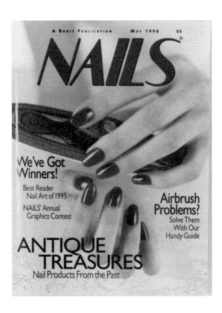

**Fig. 3-3** Nails *magazine covers the nails segment of the beauty business.*

**Fig. 3-4** Les Nouvelle Esthetiques, Skin Inc., *(shown here) and* Dermascope *provide information about the skin care, makeup, waxing, and day spa management business.*

message for any of them to be truly effective. (We'll talk about how to define your image and ensure you are presenting a consistent image in the next chapter.)

> ## Pro's Tip
>
> *You and your staff must strongly believe in your PR message. Hopefully, it's a process that you will enjoy and feel good about. The editors, writers, and broadcasters you are pitching must be able to sense and "catch" your enthusiasm about your business and about your message. Remember, if PR is not fun, then it's not worthwhile and it probably won't be effective.*

Make sure you are familiar with any publication and read it regularly before you target it. This way, you will understand the types of information those publications regularly use. In addition, if you are speaking with an editor or writer from the publication, you won't be in the dark if he or she mentions a particular section. For example, if your local newspaper runs its regular fashion and beauty section on Thursday, with a special section appearing every other Sunday, you'll want to be aware of that. Or, if your local television morning show does makeovers on the third Tuesday of every month, you'll need to know that, too.

You will also want to keep those publications that are in tune with your image in your receptionist area, with the exception of the trade magazines, which, as you know, are designed to be kept in the staff area and not put out for client view.

# A PR Plan Example

Here's an example of a PR plan that's designed to cover a broad spectrum of ideas that will all be coordinated within your salon's image. You won't do this many projects your first year out. You probably won't even do something every month. This sample plan is designed simply to give you some preliminary ideas of what we are discussing in this and subsequent chapters. It's also designed to get your creative juices flowing while we go through the important mechanical aspects of PR—such as planning and budgeting—in the first few chapters.

This sample plan will demonstrate the importance of preplanning to meet the time lines under which the press works. Your local beauty writers might work a few weeks to a few months in advance. Consumer editors are working 4 to 6 months in advance. That means if you send them a press release with your fall trend in September, you're about 4 months too late. Later in this book, we will walk you through the process of developing your own plan that's specific to your message.

As you can see by this example plan, with the exception of the Spring/Summer and Fall/Winter trend predictions, which will require a photo shoot, all of the other PR activities have little associated cost other than putting together the press release, mailing, and telephone follow-up. These are examples of how to garner press from activities you are probably already doing.

> **January**
> ◆ Write a letter to all targeted press telling them who you are and what you can offer them (Fig. 3-5). Limit it to one page only. Make it brief, professional, friendly, and informational. Also, make sure you have a professional writer proofread it and offer suggestions or have your PR agent do the letter for you.
> ◆ Start the year with personal press parties. Invite your local press in for personal tours of your salon, a consultation, and free services of their choice. (Make sure you do the services yourself, or book them with your best stylist/colorist/esthetician, and have a plan in mind for compensating the staff members who do the service. Unless they're intimately tied into your PR program so that they receive a direct benefit, it's not fair to ask them to take their time away from clients to do an editor for free.) Make sure they go home with a beautiful basket of retail products—complimentary of course!—that are specifically selected for their hair/skin/nails.
> While an editor is in your chair, you have a "captive" audience. Talk about what you do without being pushy, and ask questions about the editor's job. People love to talk about themselves, so be a good listener!

Also, if your in-salon PR is up and running (see Chapter 9), the editor will be receiving tons of positive subliminal messages.
- Call your targeted media to follow up on your letter and ask what types of articles they'll be working on or what topics they'll be researching over the next few months. (Note: If you're working with a PR agent, finding out this information is his or her job. Make sure it's being handled for you.) Take careful notes and offer to send them information. Never ask them to use you or force your information on them.

If you get a curt, "no thank you," it's best to back off for a while. But, don't get discouraged. Wait one to 2 months and try again.

**February**
- Reinforce your Spring/Summer Trend to your local press.
- Hold your Valentine's Day event.
- Send your beauty tips for brides to the local press (Fig. 3-6). Give it a specific focus: For example, create a 4-month beauty program to follow before the wedding; talk about the importance of having a hair and makeup walk through before the big day; offer tips for coping with beauty catastrophes on the wedding day; provide styling tips for honeymoon hair depending on where the honeymoon is, etc. This is just to show you that there are virtually limitless ways to approach bridal beauty. Select one that's uniquely yours.
- Release your tips for tackling summer hair problems, such as humidity, chlorine, too much sun, to national consumer press.

**March**
- Attend a national or international beauty show. Try to meet with at least one editor of the major trade magazines to talk about your salon and the editor's editorial needs. Call to set up the meeting in advance.
- Follow up with a press release to your local press to tell them that you attended, what you learned, and how you are changing your salon as a result, perhaps adding a new service or featuring a new product.
- Announce your Mother's Day programs to the local press. Include a list of beauty ideas that moms and daughters can do together for this special occasion.

**April**
- Release your tips for tackling summer hair problems, such as humidity, chlorine, too much sun, to local press.
- Release your Quick Beauty Tips for Back to School to national consumer press.
- Donate five gift certificates to a summer charity raffle. Make sure you receive credit at the raffle and in the event's program.

## 52 Planning Your PR Program

**Fig. 3-5** *If you have a largely young clientele, it pays to contact consumer magazines targeted to teens.*

**May**
- ◆ Release your Fall/Winter Trend to local, consumer, and trade media.
- ◆ Take any press you've received so far, have it matted and framed, and display it in your salon.
- ◆ Release Post-Summer Hair and Skin Care Tips to national consumer press.

**June**
- ◆ Release your Quick Beauty Tips for Back to School to local press.
- ◆ Use any down time to plan your fall media blitz.
- ◆ Invite a local editor to lunch.

**July**
- ◆ Reinforce your Fall/Winter Trend to your local press.
- ◆ Release Post-Summer Hair and Skin Care Tips to local press.

**August**
- ◆ Hold a back to school cut-a-thon to benefit a worthy cause. Encourage students going back to school to come in and get their hair cut. Release the who/what/where/when/why several weeks before the event to your local media and invite them to attend. Call local radio and TV stations and invite them to do a

Chapter Three **53**

**Fig. 3-6** *Many magazines feature special style sections for brides, so if you specialize in wedding parties, it pays to publicize that segment of your business.*

"remote" broadcast. Send a follow-up release with black and white photos to your local press, telling what happened, how much money you raised, and where it will be donated.

### September
- Release your holiday hairstyling trends and tips to local press.
- Plan special holidays events, parties, promotions, and charity participation.

### October
- Release your Spring/Summer Trend to local, national, consumer, and trade press.
- Make sure that all of your PR mentions are framed and displayed to impress all the clients who will come through your doors for the holidays. The same goes for the letters you've received that complement your work or a thank you for participating in an event.

### November
- Participate in a holiday charity event of your choice. Two to 3 weeks before the event, invite local media to attend. Do a follow-up press release announcing your participation and the results to your local press and the trade press.

**Figs. 3-7, 3-8, 3-9** *Stylist Rhonda Hicks from Houston specializes in avant garde work, so she makes sure that a portion of her press program is allocated to that creative aspect of her business.*

> **December**
> ♦ Release your Valentine's Day Beauty Treats to your local press. Feature special services that are perfect for couples. Or put together a Valentine's Day Singles' Event at your salon.
> ♦ Send holiday cards to your press list.
> ♦ Have a busy holiday season.

## Outlining Your Plan

What makes your salon more newsworthy than the salon down the block, across town, or even across the country? What do you do that's different? How have you set trends? What do you have to say or do that sets you apart from the rest?

Go back to your "What's My Image?" worksheet. The points brought out here about what makes your salon different are all good starting points, either as the focal point of a press release or as interesting background information (Figs. 3-7, 3-8, 3-9).

It might help to design your PR plan from your target audience's viewpoint. Look at your salon objectively and ask clients and staff what is unique and interesting about you and the salon. Think about the questions your clients ask you. Your answers would probably make a good informational press release. (Fig. 3-10)

If you decide to work with a PR agent, he or she will help you put your final plan together. However, it's always a good idea to have a rough plan in mind when you and your agent are discussing options. You can begin to outline your plan here. After you read the following chapters, come back to this form and fill in the gaps and time lines.

**Fig. 3-10** *If you have a good idea, let editors know. Salon owner Frances DuBose and her daughter had an innovative way to sell perms, and they told* Modern Salon *editors about it. As a result, they were asked to do a photo shoot on their technique for the magazine.*

## MY PR PLAN

| | Local | National | Industry |
|---|---|---|---|
| January | | | |
| February | | | |
| March | | | |
| April | | | |
| May | | | |
| June | | | |
| July | | | |
| August | | | |
| September | | | |
| October | | | |
| November | | | |
| December | | | |

Now that your plan is beginning to take form, always be flexible. The news media moves quickly. You must be prepared to change your plans to take advantage of what's current, just as you might change your business plan as circumstances evolve.

# Where to Send Your Message

Your PR plan will be a guideline that ensures you pitch your message to the right media, at the right times, to reach the right audience. Remember: One positive editorial in the right media is worth a thousand in the wrong ones because you'll only be attracting the wrong people. With that in mind, let's take a look at the different types of media you can target (Fig. 3-11).

All require different strategies and different levels of commitment in terms of both time and money. Keep in mind that the farther you are from your own backyard, the longer it will take to see a result and begin to achieve your goals.

**Fig. 3-11** *Modern Salon's editors are always looking for innovative stories to feature in "Clippings," a monthly section. Read the magazine regularly so that you understand what types of information they publish here.*

# Chapter 4

# Hiring and Budgeting for PR Pros

## Do You Need to Work with a PR Pro?

Would it be better for you to work with a freelance writer and then handle the PR on your own? Or would you get better results working with a PR pro from the start? If you want to work with a PR pro, then how do you go about selecting the person who is right for you? Should you look for a large agency in New York City or a private practitioner down the street?

You basically have three choices:
1. Hire a large agency to handle your PR (most costly).
2. Hire a publicist who works independently, rather than at a larger, more costly agency.
3. Hire a professional writer to edit your press releases, but do the bulk of the work yourself (most costly in terms of your time, least costly in terms of your dollars).

You need to keep several points in mind:

First, it's important to rely on a professional to write your press releases. Intriguing editors is an art and not everyone can do it, even if their spelling and grammar are perfect. Your press releases need an original concept and exciting writing that gets to the point. A three-page

**Fig. 4-1** *Florida Salon owner Chenzo Balsamo (right) has become well known by working with the Eshe & Alexander PR Agency, which has introduced him to key contacts, such as IBS's Tom Berger (left).*

release won't get read by a busy editor, no matter how important you feel the message is.

Finding writers can be more difficult than you might think because for some reason, everyone thinks that he or she can write. For example, your mother-in-law, who writes poems for her own enjoyment in her free time, might know you better than your professional writer and might do all your work for free, but that's not a good enough reason to have her write your biography. It's rare that anyone other than a pro can make your copy so inviting that everyone will want to read it. Plus, if you don't work with a professional writer, you end up with grammatical mistakes and punctuation errors, which can almost guarantee your release will get tossed in the circular file. It's similar to your clients who tell you they can cut or color their own hair as well as your staff. They might think so, but as the professional you know better.

It's fine to create "rough" copy or an outline yourself that contains all the facts about you, a service, or an event. Just make sure you have a professional edit and proofread it.

## Screening Potential PR Agents

When you screen PR people, agencies, or freelance writers, be prepared to plainly state your needs and areas of interest. You already have this in writing from filling out the forms provided in this book.

Several methods can be used to find an agency, publicist, or writer (Fig. 4-1). One of the best is to ask trade magazine editors. They will give you the names of writers and publicists with whom they like to work. This ensures that your work will get a little extra attention. Once you get the names, call them and tell them what you're looking for. Ask for samples of their work and ask to see results they have achieved for other clients.

Writers and publicists often work long distance, so you don't necessarily have to hire someone who is local. You can also ask salon own-

> **Note**
>
> *Some PR agents will only work with one salon. Others will work with only one salon per major metropolitan area. Yet others will work with salons that have different specialities. For example, they might have one salon that specializes in hair color clients, another that targets African American clients, and a third that's a major day spa. Only you can decide what you feel comfortable with. However, if you want to be an agency's only salon client in the United States, you will probably pay a considerable amount for this exclusivity.*

ers from other cities who they use. If you see their name in *Glamour* every other month, chances are they are working with a pro. If that doesn't work, ask your professional clients if they know anyone or contact professional associations or women's groups in your town.

Look for a strong writer. Ask to see examples of fashion copy the person has written. This is important because writing about hair trends and fashion is different from writing about products or people.

If your budget is small, you can contact local colleges for junior and senior PR, marketing, journalism, or communications majors. Usually, you will contact the career services, internship, or student placement department. By working with a student, you'll save on professional fees and get originality and enthusiasm. Sometimes, a student will work on your PR projects just to have a professional piece in his or her portfolio. Others are willing to trade services with you. If you trade services, specify the terms of the trade in writing, so it's clear what each party is getting and giving. Keep in mind that the Internal Revenue Service considers barters as dollar trades.

When reviewing agencies, look for someone with strong contacts with the press you are targeting. Don't hesitate to ask to see clips of what has been achieved for other clients or for references.

Next, look for personality compatibility. You will work closely with your agency account executive or publicist, so finding someone

## Pro's Tip

*If you're talking to a larger firm, find out who will actually be working with you as your account executive. It probably won't be the same person who "pitches" you for your business. You will want to talk in depth with your account executive to make sure that person understands what you want as clearly as the company's president.*

you feel comfortable with, can work well with, and have fun with is important. This relationship must be built on trust because you will be sharing intimate details about your business and future plans. Finally, look for someone who is willing to teach you about PR, rather than someone who tries to keep his or her expertise a secret. As with any hiring decision, don't hire someone just because you like the individual. The bottom line is that you want a professional who will produce results.

When you have two or three top candidates, talk to them about your goals, then ask them to develop a proposal. Their proposal will include ideas of what they will do for you for a year, including objectives to achieve, strategies, and creative tactics for getting the media's attention focused on your salon. These proposals will also include their fees.

## Making Choices

Once you choose an agency or writer, expect to be closely involved. You can't expect the agency to come up with all the ideas. You know your business best, and you know what is happening in your salon. You must communicate this information to the agency on a regular basis. Set aside a specific time each week to touch base, then fax updates throughout the week. Often, you can get so busy in the salon, you forget to tell the agency what's going on! Because timeliness is everything, you want to make sure you don't miss important opportunities. Once you're used to working together, the agency will come up with ideas that suit your purpose. Your publicist can't create news. However, he or she should be able to advise you about what is newsworthy and then give your information the proper spin to make it work. You tell your publicist what you're doing and let him or her decide how to pitch it.

Your agency will be able to guide your efforts in the right direction. What's important to you isn't always what your audience wants to know and you have to be able to look at it as an outsider, as much as that's possible.

Don't expect instant results. Be willing to work with an agency for at least a year before abandoning the relationship. National magazine credits can take 6 to 8 months to develop. Local publicity might be quicker. Achieving your objectives is the primary goal. Your business wasn't built in a week; neither will your new clippings scrapbook.

> ## Pro's Tip
>
> *A common mistake many salon owners make is to assign the PR task to the receptionist because she is "good on the phone" or a "people person." We recommend that you do not do this without first arranging additional in-depth PR training for her. If after substantial training you decide to have her handle such an important task, you will also want to compensate her for her extra effort.*

## Budgeting

PR is not "free" advertising. You pay for the time an agency or publicist spends working for you, as well as for all associated expenses even if you do it yourself, with no guarantee of ever seeing your name in print. But if you do it right, with PR, moderate spending can produce proportionally large benefits in terms of image enhancement and increased revenues (Fig. 4-2). It just may take months to see results.

The form on the facing page will help ensure that you don't overlook possibly hidden costs in your budget.

Following is an explanation/guideline for each category on the budgeting form:

### PR Agency/Professional Writer

What can you expect to spend for professional help? Everyone has a different answer (Fig. 4-3). Some salons pay only $300 to $500 per month for having a monthly trend release sent to 20 beauty editors. Others pay up to $10,000 per month to a New York City agency for

| PR Budgeting Form | | |
|---|---|---|
| Expense | Cost per Month | Yearly Cost |
| PROFESSIONAL ASSISTANCE<br>   ◆ PR agency<br>   ◆ Professional writer<br>DESIGNER<br>   ◆ Logo<br>   ◆ Letterhead<br>   ◆ Business cards<br>   ◆ Folders<br>   ◆ Other collateral materials<br>PRINTING<br>   ◆ Letterhead<br>   ◆ Folders<br>   ◆ Business cards<br>   ◆ Other collateral materials<br>PHOTO SHOOTS (2 per year)<br>   ◆ Photographer<br>   ◆ Models<br>   ◆ Makeup artist<br>   ◆ Film and processing<br>   ◆ Special printing processes<br>TRAVEL (for you and your PR agent)<br>   ◆ Airfare<br>   ◆ Hotel<br>   ◆ Taxis<br>   ◆ Food and entertainment<br>POSTAGE AND SHIPPING<br>MAILING ENVELOPES<br>TELEPHONE AND FAXING<br>MISCELLANEOUS SUPPLIES<br>   ◆ Slide holders<br>   ◆ Lunches<br>   ◆ Portfolio for presentations<br>   ◆ Gift wrap<br>   ◆<br>   ◆<br>   ◆ | | |
| **TOTAL COSTS** | | |

**Fig. 4-2** *By joining professional associations such as the National Cosmetology Association, you have access to the association's PR department, which will often help you publicize your association participation. It also offers the opportunity to participate in press worthy activities, such as the semi-annual design team, shown here. Photo courtesy of the National Cosmetology Association. 1996 Spring/Summer "ICY HOT" Design team: Women's Hair Designer Beth Hartness, Men's Hair Designer Antonio Trapani, Design Team Director Candi Ekstrom, Makeup Designer Letti Lynn, Nails Designer Paulette Agha, and Women's Hair Designer Diane Moltaji.*

unlimited distribution of their materials locally and nationally. You will probably start out somewhere in between.

There are several ways you can pay a PR professional, depending on your budget, your publicity needs, and the policies of the person you want to work with. There are advantages and drawbacks to each.

- ◆ **Monthly retainer.** This is a set fee per month. It gives you the greatest consistency and an assurance that someone is always looking out for your press needs and ensures more consistent follow-up. It also helps you adhere to your budget because it remains constant each month. Although you pay every month, whether or not you did any PR activities, you don't pay any extra in heavy PR months. Just make sure it evens out over the course of the year.

**Fig. 4-3** *A professional publicist can help position you to take advantage of media opportunities, such as a styles page in* Modern Salon, *as shown here.*

- **Per project** (pay as you go). In this case, you hire someone to write your releases and even handle the production of the press kits and mailing, but the person's responsibility is limited to a particular project. You will do more legwork on your own. You'll save during the months without activities, but pay more during more active months. You might also be charged extra for planning time. Because you know what will be spent per project, this is a much better option than paying an hourly rate, which you should try to avoid because of the lack of control over the total bill.
- **Barter.** Barter refers to trading services. This can work if you need limited assistance and trade hair services for writing services. Keep in mind that barter services are looked at as income by the International Revenue Service and must be declared as such. Of course, the corresponding expenses can be deducted.

Before you sign an agreement with a PR pro, ask the person for estimated monthly or project costs in addition to the project fee or retainer and use those to fill in your budgeting form.

## Designer

Designers are almost always paid per project or per hour. Depending on the experience of the designer, you will pay anywhere from $25 per hour to $250 or more per hour. Given that differential, it's best to agree on a project rate or a package rate if the designer is doing your entire corporate identity at once.

## Printing

Order enough materials to last for 6 months to a year, as long as you don't plan on moving the business during the year. Although your up front expenses will be greater, usually the larger the order, the less expensive the cost per piece. Get quotes from several printers and comparison shop.

## Photo Shoots

Budget for photo shoots at least twice a year for your two major seasonal trend releases.

## Travel

If you plan on targeting the national consumer magazines, you'll want to travel to New York once or twice a year to meet the beauty editors of the major publications. Your PR agent will go with you and introduce you. You will be expected to pay for travel, lodging, and miscellaneous expenses for both of you.

## Miscellaneous

The remaining categories are self-explanatory and will vary, depending on the size of your media list. Keep in mind that while you don't want to blow your budget by sending your releases all via an overnight service, you do want to make sure they are all sent via first class mail with a nice presentation.

Having a good PR plan and a rough budget in place before you get started will be invaluable to you over the course of the year. It will also help keep you on the straight and narrow in the event that you're tempted to overspend. That's not to say that you might not splurge here and there from time to time. The point is that with a plan in hand, you'll know you're splurging for the sake of achieving your goals.

# What Editors Want:
## A Freelance Writer's Point of View

Freelancers who write for dozens of magazines need more sources than the average writer. After all, it wouldn't be appropriate to use the same source for three different magazines. Editors expect something new, original, and exclusive.

Because freelance writers are always look for something new, they're often easier to approach than full-time staff writers (Fig. 4-4). Without a steady supply of new information, we wouldn't make a living and unlike staff writers, we don't get hundreds of press kits a week.

**What freelance writers do:** Most freelance writers pitch magazines on their own story ideas, so if you have an idea that you think would make a good article, it makes sense to share it with a freelance writer, who will shape it for an appropriate publication, pitch it for you, and use you as a source. Just don't give the same idea to more

**Fig. 4-4** *It pays to build a relationship with freelance writers. For example, Victoria Wurdinger writes both style and business articles for a variety of consumer and trade publications, including* Celebrity Hairstyles, Hairdo, *and* Modern Salon *(shown here).*

than one writer. It would be useless to pitch a magazine on an idea that some other writer suggested the day before or that the editor already rejected.

Because I write for many hair and beauty magazines, I'm usually working on 3 to 4 different features at once. So, I need everything from new trends and techniques to business tips. For instance, one story I'm writing might be on adding relaxing services to your salon; another feature might be on how to sell add-on services; another still might be a piece for a consumer magazine on how to find a good hairdresser.

**What I'm looking for:** New, simple, usable tips are always welcome. In addition, because I write for both trade and consumer magazines, I might use a good idea twice, because there's no conflict between the two. For example, if you share a new hair coloring technique with me, I might describe it in detail for a trade magazine article on hair color techniques and mention it in a consumer magazine article on new color trends.

**How to find freelancers:** The best way to find freelancers is to look at bylines in magazines and see if the person is on the masthead. Sometimes, freelancers are listed as contributing writers or editors. You'll recognize names after a while. Call the magazine and ask how you can contact the person. Usually the magazine won't give out writers' telephone numbers, but they will pass on messages and freelance writers always return calls. It's nice to hear from someone who reads your work, and we're always looking to add to our network of sources.

**Other pointers:**
- Do add us to your mailing list. We appreciate your press kits far more than a deluged magazine writer because we get fewer of them.
- If you traveled to a show overseas and saw something new, call and share it. This is the sort of information that's harder for a freelance writer to get and is well appreciated.
- Never tell competitors what a writer is working on. Dozens of times, I've had sources tell me what someone else interviewed them about. I can only assume they're telling others what I'm writing, before it's published.
- Keep in mind that many freelance writers work out of home offices. We don't appreciate business calls at 8 AM or in the middle of dinner but somehow, we still get them. Limit contact to office hours, unless you've been invited to call at other times.

◆ Be as professional as you would if you were speaking to a staff writer. For some reason, just because freelancers don't represent one particular magazine, callers tend to get more personal and loose. I've had hairdressers do everything from make racist remarks to me to say they cheat on their taxes.

*— Victoria Wurdinger*
*Freelance Writer*

# Chapter 5

# Creating a Name in Your Own Backyard

*T*hus far you have defined who you are and what type of clientele you want to attract. You've also learned some general strategies for targeting the media. Now, you're ready for the first step toward getting yourself some public exposure (Fig. 5-1). The first and most logical place to start is locally, in your neighborhood and city.

**Fig. 5-1** *Consistency pays. Stylist Edwin Fontanez has been named "Best Stylist" and featured in* Cleveland Magazine *several times because of his continuous PR efforts.*

**Fig. 5-2** Day Spa pioneer Noel de Caprio is interviewed by Sara Arnold of Channel 12 news during the grand reopening of Noëlle Spa for Beauty & Wellness in Stamford, CT. Photo by Janet Durrans.

If one goal of your press is to create greater awareness of your salon in the eyes of potential clients, there's no better place to be seen than in your own backyard. Local PR can focus on all of the media around you, including daily, weekly, special interest, and school newspapers; radio programs; and network and cable TV programs. It includes getting credit for doing the hair for your newspaper's fashion section, defining a new trend you picked up on a trip in an article for the fashion section, or being seen in the society section of the paper at a local charity function. Local PR also includes all of the events you do, such as salon events, donating gift certificates to charity raffles, and doing the hair for local fashion shows. Anything that draws attention to you can be considered PR (Fig. 5-2).

# Your Event Checklist and Budget Form

Use this checklist and budget form whenever you hold an event at your salon or participate in an event with another organization, such as a women's group's charity event or fashion show outside your salon. Remember, even if you don't have to pay to participate, you will always make an investment in time and manpower, so you have to carefully decide which events to go for and which to refuse (see Chapter 8 for more on special events).

---

Time Commitment
    Planning meetings
    Ticket selling
    Promoting
    Rehearsals
    Actual event

Dollar Investment
    Donation to event
    Time out of salon
    Decorations
    Costumes
    Refreshments
    Products

---

# Why Local PR?

Some salons make the mistake of overlooking the importance of local PR because they want to make a national name for themselves. If that's your goal, then focus your efforts on it. But keep in mind that local and national PR programs run together can continuously feed one another and ensure that you receive the best overall results for the effort in both places (Figs. 5-3, 5-4).

    Obviously, one great benefit of local PR is that it's right in front of potential clients. If they see your name frequently, they might eventually be intrigued enough to stop by and see what you're all about. In

**Figs. 5-3, 5-4** *Stylist Edwin Fontanez receives tons of calls from new clients when his styles, like these, appear in the* Cleveland Plain Dealer's *style section.*

addition, a successful local PR program ensures that good things are constantly being said about you in front of your staff. The pride it instills in them will help keep them excited about working at your salon. Invite them to participate in your PR events and the enthusiasm grows even stronger. Another benefit is that local PR places your salon in a great light in front of potential employees. If you could fill a few more stations but just can't seem to find the right people, a strong local PR program might bring potential employees calling at your door.

## Beginning a Local PR Program

To begin your local PR program, understand your own image and goals. Next, put together your media list. Do you want to target publications only or are you interested in appearing on radio and TV, as well? How would your like to receive credit in the program or even a moment in the spotlight yourself for doing the hair for a local fashion show? You can even hold events at your own salon (Fig. 5-5).

You might want to start with print until you're really comfortable presenting your story, then progress to radio and TV. If you just aren't comfortable in front of a camera, then either assign one of your trusted

**Fig. 5-5** *Stylist Chenzo Balsamo is often asked to participate in local photo shoots.*

staff members to be the on-camera demo person or try to pitch a segment that includes a group of people for your first time.

Once you decide which types of media you want to target, narrow it down further to the publications your clients and people you want to attract as clients are likely to read, radio stations they like to listen to, and TV programs they like to watch. For example, if your clients tend to be the society women you would probably target very different media than a salon who services primarily teens.

Now, find out who to send your release to. For a newspaper, that person would be the fashion and beauty editor or the lifestyle editor. For a smaller neighborhood newspaper, you might send your materials to the features editor, news editor, or even the editor. Call to make sure you have the right name, spelled correctly.

For TV and radio, call the station and ask who the best person is to send your releases to, usually a producer or associate producer.

Now, find out how your contact person likes to receive your pitch. Some will accept only typewritten press releases or letters; others prefer phone calls. Some want releases faxed. Others will not accept faxes. Find out what days are best to call. If a weekly neighborhood magazine comes out every Thursday, then they usually won't take calls on Tuesday or Wednesday when they're closing the issue, but they might be happy to talk with you on Friday.

Next, send a letter to each of the contact names along with your backgrounder. Invite them in for a complimentary visit for any of your services. This is by far the best way to get to know your local press and show them what you and your salon are all about. There's no better testimonial to how wonderful your services are than when they experience them first hand.

Be sure to include the name and direct phone number of the person at your salon—probably you or your receptionist—who will act as your press liaison. Editors will reach someone who knows who they are and greets them appropriately when they call to set up an appointment.

### Press Discount Policy

Many media people and editors appreciate and often expect a discount on salon services. Others are not permitted to accept any service or product freebies because of company policy. To avoid confusion and any potentially embarrassing scenes at reception, make sure you have a clear press discount policy in writing. Make sure everyone knows what it is and sticks to it.

A good rule of thumb is to offer the first visit completely on the house, then you can determine what discount this person should receive in the future. Your press discounts can range anywhere from 25% to 100% off all services, but keep in mind that editors talk. If you offer one a 100% discount and another less, expect them to know.

Provided you can back up what you say, once you get your PR program rolling and as you continue to network, you will learn about more opportunities that you would have never guessed existed. And you'll be positioned to take advantage of them.

### Pro's Tip

*Look at offering complimentary services to the press—not as a bribe that they'll repay. The idea is that by coming to your salon, they'll learn first hand what you're all about and why you are press worthy.*

## How to Let People Know You're the Best: Edwin Fontanez

Edwin Fontanez was recently named Best Stylist in the City by *Cleveland Magazine*, and he was as surprised as anyone. It's this modest approach taken by this talented young hairstylist that has garnered him tons of accolades and opportunities since he began to promote himself on the fashion scene.

"You can be the best stylist in town, but if you don't promote yourself, no one but your clients will ever know," says Edwin, who does the hair for about 10 fashion shoots per year for the *Cleveland Plain Dealer*, the area's major newspaper.

That's thanks to the close relationship he's developed over the years with Fashion Editor Janet McCue. He met Janet when she came in to Dino Palmieri salon where he works and Edwin did her hair. She loved it and came back. As they got to know each other, Edwin began to talk with Janet about designers, trends, photographers, and models they both liked. Once they had developed a comfortable rapport, he showed her his work and asked her opinion. She liked what she saw, and the rest, as they say, is history.

Edwin began his climb through the fashion ranks by connecting with local models. He would be at the clubs they liked and meet them when the opportunity presented itself. Because he was so up to date on all aspects of fashion from clothing to photographers, he was able to talk to them immediately. Then he would begin to tell them about himself and eventually offer to do their hair for free.

"This business is 90% word of mouth when it comes to promoting yourself," says Edwin. "When the models liked their hair, they would tell all of their model friends and my reputation began to grow. Before I knew it, they were asking if I could help out on photo shoots. Then it snowballed and the model agencies took me on as a stylist."

In addition, he continued to send his bio and a trend release out to anyone who would take them and could help him.

Today, Edwin's relationships and hard work have paid off nationally, too. Last fall, Janet opened the doors so he could assist hairdressers backstage at the Carolina Herrera and Oscar de la Renta fashion shows in New York City. Of course, he did his best work and got to know the big name hairdressers, watching and learning. As a result, for the spring collections, he was asked back to play an even bigger role.

If you want to make it locally, Edwin offers these tips:
- ◆ Be seen at fashion events and keep up on every aspect of the business to keep your press moving in the right direction.
- ◆ Stay well rounded so that you can offer the press many services.
- ◆ Don't be pushy. In a business that's dominated by egos, modesty, without underplaying your talent, truly is the best policy. Be 100% genuine; be confident in your work and it will pay off.

**Fig. 5-6** *Frank Alvarez*

## Looking Good with Frank Alvarez

Salon owner and stylist Frank Alvarez has handled his own PR for 25 years, and it's still going strong on local TV and radio (Fig. 5-6).

He began his rise to media stardom by doing makeovers on a local TV program. When he was seen there, he was invited to host his own radio show and answer listeners' beauty and fashion questions on the air. He called his first show, "Looking Good," and that's the name he's stuck with ever since.

His stints have included everything from the man with the microphone interviewing people about fashion and beauty for *PM Magazine* to presenting the newest trends in hair, fashion, makeup, sunglasses, and other accessories.

"The key to the first and every success is persistence," says Frank. "If you want PR, you have to go out and get it." Here are his tips for getting a foot in the door:

- Always have a press kit that looks classy and professional.
- Continually mail your photo shoots, trend predictions, and tips to your media at least four times a year. The more they see you, the more they'll think about you when they need a source or an on-air personality.

"They might not call you the first time, but if you're persistent in a nice way, you'll eventually get their attention," advises Frank.

- Be fashionable or flamboyant enough that you appear to be a fashion expert without going overboard.
- Turn everything you do into a PR opportunity.

  "I've discovered that the best way to get media in your own backyard, especially if you're in a smaller market, is to show a connection to the outside," says Frank. "For example, I continually travel to New York and Europe. Then, when I come back, I let all of the media know where I've gone and what I've seen. If I can't get a response from the beauty or fashion editor, I'll try the gossip editor. If I keep trying, someone will cover it."

  In fact, the day after we talked with Frank, he was leaving for a trip to Cuba. He had already planned how to pitch what he saw there in terms of beauty and fashion to the local media the week after he returned.

- Good PR is good for business.

  "When clients and potential clients see you on TV, they automatically think you are good. We all tend to be impressed by 'celebrities,' even local ones. That's credibility that no amount of money can buy. They don't realize that you're a person who is just like them. The difference is that you knew how to get on TV."

- Stage presence comes from knowing your topic inside and out.
- You don't have to spend a lot of money to be on TV. In fact, once Frank became known, the station paid him for his appearances.

To support your publicity program and help maximize your exposure locally, follow these pointers:

- **Be seen.** Make a point for you or a trusted representative of your salon to attend those events that you would like to participate in and those where all the media types tend to hang out (Fig. 5-7). Anywhere from women's groups to particular restaurants to clubs is appropriate.
- **Network.** Get involved with the local civic organizations that your clients belong to. This is another time when it's beneficial to have your staff involved because they can help you divide and conquer if your clientele is somewhat diverse. Your younger stylists can frequent the clubs and the coffee bars while you can tackle the more professionally focused groups.
- **Go where the fashion goes.** Find out where models, photographers, fashion stylists, and other fashion types socialize. You go there, too. Connections made socially tend to pay off in business.

**Fig. 5-7** *Even a catastrophe can be turned into good press. Stamford, CT Mayor Stanley Esposito and Michael Macri, department of public works, congratulate Noel and Peter de Caprio on the grand opening of Noëlle Spa for Beauty & Wellness after fire destroyed the original location. Photo: Janet Durrans.*

**Fig. 5-8** *Good deeds come back to you. Stamford, CT held a parade complete with balloons and banners to congratulate Noel and Peter de Caprio. Photo: Janet Durrans.*

# Chapter 6

# Making a Name Nationally

To make a name nationally, you will target popular consumer magazines, such as *Glamour, Vogue, Mademoiselle, Elle,* and others. A national PR campaign requires a sizeable investment of time, money, and energy. Although these mentions are not easy to come by—most will not publish a salon's hair trend photographs because they shoot their own materials—they are possible and even likely when you approach them professionally and with an understanding of what these editors want or don't want. That is where a professional PR agency or publicist can be most helpful. These people work with the media full-time and have the experience and contacts needed to get you in the door.

The first step is to decide what kinds of magazines you want to target and then put your list together. (One advantage of working with a publicist here is that he or she will already have a list and you should expect that person to keep a current list for you.) Otherwise, you can put your own national press list together or purchase one. We have included a list of some of the most popular titles in this book. Several companies also sell media directories that come out for the next year in the late fall. Two of the most popular services are called Bacon's Information Inc. 1–(800)–972–9252 and Burrelle's 1–(800)–876–3342. However, know that it will cost you a few hundred dollars. No matter which route you take, it is your responsibility to update the list every month by checking the masthead when each magazine comes out.

Once your list is together, you begin your campaign by sending each magazine a targeted backgrounder kit. You may delete certain materials and add others, based on the magazine's readership. For example, the tips you would include in a press kit for *Vogue* (a high

**Fig. 6-1** *National consumer magazines such as* Marie Claire, Good Housekeeping *(shown here),* Self, *and* Essence *regularly provide their readers with beauty tips they can use at home. They love to receive tips from salons around the country that they can share.*

fashion magazine) would be very different from those you would send to *Teen* (a service magazine targeted specifically to teens) or *Good Housekeeping* (a service publication with a huge circulation)(Fig. 6-1).

Although most consumer editors do want to hear from you, do not become a pest. If you really must call to see if they received your material, ask to speak with the editor's assistant, then keep your call brief and to the point.

Once you have sent your backgrounder and several other releases and have received some positive response, you might want to try planning a press trip to New York City to actually meet with editors. Again, while this is done most easily with the assistance of a publicist, it is not impossible to do on your own. Some points to keep in mind:

- Have a reason to request a meeting with an editor—a new trend or a new service. Just to say "hi" is not good enough to take up an editor's time.
- Call and tell the editor you will be in New York City and ask if you can stop by for a short time to present your story and meet the editor or assistant. Don't take a "no" personally. The editors are always busy and a refusal usually reflects a lack of meeting time.

- Bring a professional presentation that completely gives the details you want to present, as well as any visuals that will aid your presentation, for example, a product, a photo, your portfolio or styles...whatever you think you will need.
- Practice your presentation many times with different target audiences before going to New York.
- Always arrive on time and leave yourself enough time before the next meeting in case the editor is busy and you must wait for a few minutes.
- If you aren't familiar with New York City, find someone who is to help you set up your schedule, so you can try to schedule nearby magazines close together in your schedule.
- For your first trip, try to see at most 10 magazines—fewer is better.

If you do get in the door, these meetings can run anywhere from 5 to 20 minutes. Just keep in mind that you are there to provide information an editor needs, not to tell her everything you think is important.

One fairly new benefit of connecting with consumer magazines is that many are expanding the amount of information they offer to their readers by setting up sites online. One example is *Seventeen*, which has its own site on America Online.

According to Simon Dumenco, executive editor of *Seventeen* Online, "The number of pages available for editorial in magazines will continue to shrink. But, the space available online is really unlimited. That's why we are taking advantage of this new opportunity to serve the readers of *Seventeen*, while attracting new readers online. At the same time, we can use more of the information that is sent to us."

What are they looking for? "Beauty tips," says Sophie Knight, Assistant Beauty Editor for the magazine and a regular contributor to *Seventeen* Online. "We love tips."

In the following pages, we have talked with editors from a variety of fashion, beauty, and services magazines to let you know how to work with them best.

## Meet Lenny LaCour

Chicago hairstylist Lenny LaCour has made quite a name for himself (Fig. 6-2). In fact, in the past year, he spoke to Miss America pageant contestants about hair care and hair color and appeared on the TV broadcast and did hair at the prestigious Seventh on Sixth fashion shows. How did he do it? Through his link with Clairol as a PR consultant.

**Figs. 6-2** *Lenny LaCour*

Lenny met Clairol's staff many years ago through mutual friends and immediately formed a close relationship. As a result, over the years, they have booked him for everything from the above prestigious national events to stage presentations at national trade shows to press parties. Many companies are willing to hire outside consultants when they have an event to staff. Two criteria: Lenny does have to commit his own time for these activities, which means time away from behind the chair, and he has to do his own research for each appearance.

In addition, thanks to continuous networking, Lenny has styled the hair for both the Democratic and Republican conventions and even the Inaugural Ball.

While Lenny received initial press for participating in these events, he took it one step further on his own to get a second hit with the media. He contacted his local press and received even more coverage about his hobnobbing on the society and gossip pages of his local papers.

What's next? Lenny has developed his own rollers, which he is marketing through a variety of media. Using what he's learned about the press, he put together a publicity campaign for his new rollers that earned him mentions in every magazine from *Mademoiselle* to *Teen* to *Ladies Home Journal*. In addition, years of relationship building has paid off in the form of his own regular segment on a local morning television news program.

# National Publications: What Editors Want

To give you an idea of what some top editors want to know from you, we asked them. Here are their answers.

## Good Housekeeping

**How to contact us:** We enjoy talking with salons around the country, so you're welcome to call if you have information to share. It's a good idea to send background before calling so we have something to refer to. A short bio about you or your salon and a list of salon services or your menu is enough.

**What to send:** We love photographs of techniques that are new, unique, and exciting, for example, a new two-step highlighting. We also like anything that's newsy and tips that are unique and different.

**The key to being heard:** Offer something to us. We get more than 100 press kits a week. You can stand out by sharing something that's unique, above and beyond generic salon news. If you see a trend in your region, we'd love to know about it. Be prepared to tell us what the trend is, where it's coming from, and where it's going. Also, anything interesting you can share about your clients is great. For example, what are your clients' most common problems and how do you solve them?

Don't call and ask what we're working on. We'd rather hear about what you're working on.

– *Karyn Repinski*
*Beauty Editor*

## Self

**What we're looking for:** We like to hear from salons. We're looking for stylists who are innovative, yet still realistic with ideas that are timely. Keep in mind that our lead time is about 4 months.

**How to contact us:** The best way to contact *Self* is to send an introductory letter with samples of what you have done. Tear sheets of what you've had published or current photos or photocopies of your new trend are good places to start. Include a brief description of your salon's philosophy and who your clients are.

It doesn't matter if you have a publicist, as long as your presentation is professional and keeps our needs in mind.

One call to follow up is fine. Just please don't continue to call constantly.

– *Maureen Meltzer McGrath*
*Beauty News Editor*

## Marie Claire

**What we're all about:** *Marie Claire* is a true blend of a fashion/beauty and service publication. That means the magazine contains both trend information and handy tips.

**How you can work with us:** We like to hear about trends from all around the country. Although we might not agree that your vision is a trend, we like to talk with stylists who see themselves as trend interpreters. We also like to see your trends if you have photos or slides. Beyond that, tell us about you and your clientele.

Tips that readers can use at home are big winners. We do quite a few articles that are service oriented and are always looking for great information. For example, sending us three great tips for getting the perfect updo in 3 minutes would be useful.

**Stay in touch:** It's good to make one follow-up call after you send a press kit because it will remind us that you'd love to be a source. Plus, if we're working on a story when you call, we might include you right away. We have an extensive filing system and will keep your information, so you may even get a call 6 months down the road. Don't call to ask what we're working on. Instead, call and tell us that you'd like to be a resource and what you have to offer.

To show us that you're serious, it pays to be consistent with your PR effort. Sending us information every 2 months is frequent enough to keep our attention.

*– Alexandra Parnass*
*Beauty and Fitness Director*

## Redbook

**What we're looking for:** Tip and trend information from all over the country. Although *Redbook* features both tips and trends, we only run trends that offer something to the reader. We don't run trends for the sake of trend reporting because we are primarily a service magazine.

**How to get our attention:** Give us an exclusive and let us know that this tip or story is only ours. We have such a long lead time that if you send us the same information you send to other magazines, we probably won't run it because we would be several months behind. We also love to have new sources contact us.

**When to call:** Pick up the phone only if you have something very specific to say. It can be a tip, a trend, a tippy trend, or a specific story idea. Just please keep in mind that I'm always swamped with calls and press kits. Don't call to follow up unless I've asked you to call me. Also, don't send slides or photos. It's a waste of your money because we won't use them.

*– Marcia Menter*
*Beauty and Style Editor*

# Hairstyling Publications

Consumer hairstyling publications are found in grocery stores, drug stores, mass merchandisers, on newsstands, and in a variety of other outlets. Many salons even keep some of these publications in their reception areas so their clients can use them as style selectors.

These hairstyling publications are wonderful outlets for your photos. They use many photos in each issue, so as long as you meet their needs, they're likely to use your work again and again. They love to develop close relationships with hairstylists across the country, so chances are, they'll be open to your calls. These companies publish many different titles, several times a year. They reach tens of millions of consumers, which means you can become "a name" quickly.

Developing a relationship with the editors of these publications can pay off with the most press clippings of any of your efforts because each editor oversees more than 10 publications per year. Tailor your photo shoot and information to what these magazines want and you could be published over and over again.

Chances are, your clients will see these publications. Just in case, don't forget to fill your reception area with them.

# What Editors Want

## Harris Publications

**Who we are:** Harris Publications publishes 10 different titles several times a year, including *Celebrity Hair Styles, Short Hair, HairDo Ideas, Soap Star Hairdos,* and more. These are found around the world in newsstand and magazine outlets.

**What we look for:** The most important thing is really good photography. Although many people think they can take their own hairstyle photos, the truth is that this rarely works. It pays to hire a professional.

We like straightforward poses in front of a white, seamless background. A pretty, All-American model with pretty hair and pretty makeup works best for us. It's a good idea to hire a professional makeup artist, as well. Often, we receive photos in which the hair is really great, but the makeup is gaudy, so we can't use them.

Send color slides or transparencies or black and white photos. Include all credits you would like run with your photo shoot and signed releases for every model.

**What we love:** Makeovers. Many different looks from one haircut on a model.

**Want extra consideration?** Then send three sets of each press kit. That way, I can consider it for more than one magazine at a time.

Please don't call and ask when and if your photos will appear. If you work is used, I will send you a magazine eventually if you provide your address.

– Mary Greenberg
Editor

## GCR Publications

**What we are about:** At GCR, we publish more than 20 magazines a year. Just some of our titles include *Short Hair Styles, Step-by-Step Hairstyling* (Fig 6-3), *Instant Hair Styles, Complete Short Hair Styling Guide, Complete Hair and Beauty Guide, Hairdo and Makeover,* and *101 New Hairstyle Ideas*.

**Fig. 6-3** *A hairstyle magazine from GCR*

**What we like to receive:** Because we have so many magazines, we can always use new material from hairstylists. We like simple, commercial styles with plain backgrounds.

**How to send us your work:** Send us either 8" x 10" black and white glossy photos or color slides/transparencies that are protected with slide holders and cardboard. We also accept slides/transparencies for cover consideration. To be considered for a cover, a look must be in color and show a simple, commercial hairstyle with a solid background. We don't often use avant garde work.

With your photos, send any information that you would like us to include, such as your name, your salon's name, the photographer's name, makeup credits, and so forth.

Feel free to call me if you have any questions. I love to hear from you!

*– Sandy Kosherick*
*Managing Editor*

## *Sophisticate's Hairstyle Guide* and *Sophisticate's Black Hairstyle Guide*

**What I'm looking for:** The information I desire from salons is nice, descriptive copy regarding the release along with styling instructions (Fig. 6-4). Photos can be submitted as black and white prints, color prints, or color slides/transparencies. Lighting should be good and clear and styles should be wearable. Choose models who are appealing. Show as many views as possible, including a front view, profile, and back view. Always include model and photographer releases.

We also contact salons who tell us they are interested in shoots and keep their information on record.

**Do you need a publicist?** I don't have a preference if a salon works with or without a publicist...it sometimes simply depends on how well versed a salon owner is in conducting photos sessions. Publicists I've had long relationships with are usually aware of the quality we are looking for and can deliver it so the work is more likely to be published.

**Pet Peeves:** What I dislike the most is receiving really great photos with no model or photographer releases. When that happens, everyone loses out.

*– Bonnie Krueger*
*Editor-in-Chief*
*Sophisticate's Hairstyle Guides and*
*Sophisticate's Black Hairstyles & Care Guide*

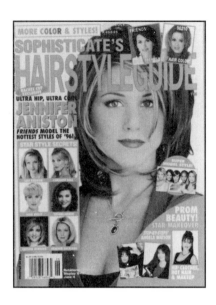

**Fig. 6-4** Sophisticate's Hairstyle Guide *frequently features tips from the stylists to the stars.*

# National Television

Although it's difficult, it's not totally impossible to do hair for a national TV show (Fig. 6-5). If you live in a major media network, such as New York, Chicago, or Los Angeles, your chances are probably better of working with one of these shows.

Often, connections are made when someone from the show visits a salon to have his or her hair or nails done or knows someone else who goes to the salon and raves about it. (Just one more reason it pays to know who your clients are.)

Talk shows and news/feature shows are your best bets. Many popular national TV shows, such as *Jenny Jones*, do "theme" makeover shows almost weekly. Although they do tend to work with hairdressers in their local area, it is possible to form a relationship that works for you.

Again, first make a list of shows you are interested in "pitching." Next, call and find out who the producers or segment producers are. Then, watch those shows for a couple of weeks to make sure they are in line with your image. For example, if your clients all tend to be conservative, Bible Belt clients, you probably won't want to be involved in a segment on college lesbian makeovers.

**Fig. 6-5** *Stylist Mary Brunetti receives publicity as a result of styling talk show hostess Ricki Lake's hair.*

The good news here is that once you do work for one of these shows and they like you and your work, they'll tend to ask you back. For example, the artistic team from the Chicago-based Mario Tricoci Hair Salons and Day Spas did the *Jenny Jones* show eight times and *Oprah* (both also Chicago-based shows) four times in one year.

# Chapter 7

# Getting Known in Your Industry

*P*rofessional publications you read every month, such as *Modern Salon, Salon News, SalonOvations* and others, are known as the trade press. They service the readers of a particular profession or industry.

You will want to focus your resources on a trade PR program if you desire greater recognition and respect from your peers. You want to be seen as a leader of your profession. You want manufacturers and distributors to notice you. You'll also want to make a name in the trades if you're considering inventing a tool or product that you eventually want to sell to other hairdressers. Trade PR is important if your long-term goal is to do work for a manufacturer as an educator, artistic team member, or spokesperson. Finally, trade PR is the medium that allows you to give back to your profession by providing information that can help other beauty professionals be more successful.

Trade PR can also have a positive impact on your other PR efforts (Fig. 7-1). How? Many of the consumer magazine editors read or at least skim the trade magazines to see what's new and who's on top of the trends. In addition, if you're quoted or your hairstyle is featured in *Modern Salon*, you've won bragging rights to your clients and will really impress them. You can also include trade tear sheets in your media kit that you send to your local and consumer magazines to show them that you are respected by your peers.

Many of the trade publications have similar needs in regard to the quality of photos, release forms, information, and complete credits. Beyond that, they all have their own unique viewpoint based on their target audience and readership. It's important to be well aware of their differences before pitching an idea. Following are examples of what many of the leading publications are looking for.

**Fig. 7-1** *Hairstylists who work part-time as educators for manufacturers can usually tap the company's publicist to help them get press. Here, Dennis Mitchell, an ABBA Master Affiliate from Matie Salon in Ventura, CA, and his model pose for press photos after an ABBA educational event. The photo was subsequently published by a number of trade magazines.*

### Pro's Tip

You won't want to send local publications press clips that appeared in the competition, so trade press will provide another source of tear sheets.

**Fig. 7-2** Modern Salon *is a national magazine that taps hundreds of beauty professionals a year as sources. Will you be next?*

# Trade Publications: What Editors Want

## *Modern Salon*

Although *Modern Salon* is based in Lincolnshire, IL, this popular trade magazine (Fig. 7-2) also has senior staff members on the East and West Coasts to ensure thorough coverage of salons and markets across the country.

**What we are about:** *Modern Salon's* focus is education—any information that enhances the creativity, technical ability, and business acumen of the salon (Figs. 7-3, 7-4). That is the type of information that we try to provide to our readers on a regular basis, and that is what we encourage salons to share with their colleagues.

Most of our material is staff created and staff produced, so it's best not to spend time and money on photo shoots or stories without obtaining a specific assignment from a member of the *Modern Salon* staff.

**100** *Getting Known in Your Industry*

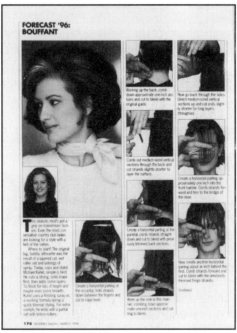

**Figs. 7-3, 7-4** *Keeping your pitch educational is the key to earning placement in* Modern Salon. *This is an example of a step-by-step shoot for a new technique.*

One exception to this is photo collections, which we occasionally publish from unsolicited submissions. These collections should be designed around a unifying theme, such as short hair, bangs, or spring color trends. Photography, makeup, and wardrobe should also be consistent. Use professional photographers, models, and makeup artists if at all possible.

**How to build a relationship:** The key to public relations is *relationship* and relationships are built over time. The best way for salons to approach me is with an introductory phone call, followed up with a press kit containing information about the salon, samples of their work, press clippings, and any data they wish to share about specific techniques or business practices. Then, be patient and stay in touch. If we can't use something immediately, we may be able to use it in several months. Or, we may contact you about something completely different in the future, depending on what we are working on. At some point, you might be invited to contribute as a source to a story or to take part in a photo shoot.

**Do you need a publicist?** It is irrelevant whether a salon works with a publicist or not as long as you are courteous, patient, and professional and your materials are well presented.

**Pet Peeve:** Often, people contact me and say, "I'm calling because I think this should be in your magazine." My question is usually, "Why?" and often the answer shows no regard for our content or to our readers. Please keep in mind that *Modern Salon* is a *service* publication and that the information we publish is not meant to serve us or our sources, but our readers. If you keep that in mind when pitching your ideas, you'll have a better chance of being received positively.

**Some final pointers:**
- Submit four-color transparencies or black and white photographs. Color prints will not be accepted.
- Be sure your name, salon name, address, and telephone number appear on all slides or photographs.
- Include some general information on the techniques used to achieve the styles and/or the overall concept behind the images.
- Strive to create thoughtful photographic collections (three or more images) designed around a particular theme (i.e., Short Hair, Spring Blondes, Romanticism, etc.).
- Unless you are striving for an avant garde approach, styles should be clean, commercial, and wearable.

- Use professional models, photographers, and makeup artists.
- Include complete photo credits—the name of the photographer, makeup artists, hair stylists, perm or color technician, and fashion stylist, if applicable.
- Enclose model and photographer releases that extend full rights to you and anyone you choose for reproduction.
- Issues are completed at least 2 months prior to an issue date, so expect to see material published (if accepted) anywhere from 2 to 6 months after you mail it.
- Obtain a copy of the current editorial calendar for specific issue themes and articles.

*– Jackie Summers*
*Editor in Chief*
*Lincolnshire, IL.*

## ... and More Tips from the East Coast

- I work exclusively with salons in the eastern United States.
- Send information regularly. Whether it's four, three, two times a year, or even simply a yearly trend prediction, the more materials you send, the better chance I have of recognizing your name when I see it again (Fig. 7-5).
- "Clippings" is our newsy section at the front of the magazine where we feature a variety of stories. We especially love people stories, such as a salon that does a cut-a-thon or has a unique idea. Include more information than we need and the answers to who, what, when, where, why. Send slides or photos to illustrate your story whenever possible.
- Make sure your model is as pretty as possible from all angles.
- Include your salon name, a contact name, and telephone number on each page of the press release.
- I like to take notes on the press folders. You'll make it easy for me if you send your press materials in a light-colored folder made of a material I can write on.
- Don't send technicals. We produce our own.
- We love to receive men's styles because we get so few of them.
- Check the spellings of your credits carefully. We assume that what you send in is correct and will not call to double check.

**Fig. 7-5** *Hairstylists who are selected to work on* Modern Salon's *photo shoots are profiled in the front of the magazine every month.*

- If you want to know if we received your materials, send them certified. If you must call to follow up, ask for my assistant.
- Do not ask for your materials back.
- Don't expect to see your materials in the magazine for at least 3 months, often longer. But, if it never runs, don't feel bad. Your information might just not have fit into a particular issue. Try again.
- If your goal is to be involved with a photo shoot, send pictures of the kind of work you do.
- If you don't hire a publicist, then designate a particular person in the salon who is committed to your PR program. Don't just give it to the receptionist. When the salon takes this step and gives me a regular contact, I know the owner is committed to the publicity program.
- Additional publications (Figs. 7-6, 7-7).

*– Maggie Mulhern*
*Beauty Editor*
*New York City*

**104** *Getting Known in Your Industry*

**Fig. 7-6, 7-7** *A sister publication of* Modern Salon, Salon Today *targets salon owners and features business related issues that offer advice and case studies for successfully running a salon.*

# *Salon News*

**What we are about:** Salon owners are the number one priority of *Salon News* editors. Indeed, we would not be here without the participation of our readers. We are dedicated to delivering the information they need to grow and prosper, and to making sure that a wide range of opinions, experiences, and geographical locations are represented. My goal is to, at some point, include coverage—a quote, a photo, etc.—from every single one of our 75,000 readers.

What makes *Salon News* unique is the fact that we direct our editorial specifically at owners of full-service salons. All of our coverage—fashion and beauty trends, features on management, marketing, technology, and retail issues, and departments, such as Happenings, Update and Style Alert—is created with owners in mind. The one specific editorial area we do not cover is technicals; thus, there's no point in sending detailed photographs or minute descriptions of specific details.

**PR Opportunities:** Our editorial mix has three parts: beauty/fashion trends, salon management information, and news. We invite salon owner input in all of these areas. We also welcome targeted queries for the following departments:

- "Happenings," which showcases the latest events and achievements, including trade shows, seminars, salon openings, and charity efforts
- "Update," which delivers short briefs on news and upcoming events
- "Hip Clips," which is our showcase for salon creative work. This section showcases current work and requires photos accompanied by a thorough explanation of the concept and technique used.

For all of these, call, write, or fax us. Submissions of color slides/transparencies or black and white prints are welcome for both Happenings and Style Alert.

For visuals, we prefer color slides or transparencies, although good black and white prints are acceptable, too. Every piece should be clearly marked with the salon's name, city and telephone number, as well as with identifying information (i.e., a caption on photos). In the case of illustrations, salons should send slides. (That means the illustrations will be photographed or output from the computer in slide form.) Original artwork is often damaged in shipping. Unless specifically requested, we cannot return materials.

**Deadlines:** In general, the deadline for material to be included in a particular issue is 6 weeks preceding the first day of the month of publication. For example, the deadline for the May issue is March 15. However, exceptions are made to accommodate late-breaking, time-sensitive news.

**How to contact us:** Any means of contact is welcome, be it a telephone call, press release, letter, fax, trade show conversation or other. Because we have such a small staff, we depend on our readers to get in touch with us. We also identify our salon owner sources through lists provided by industry organizations, manufacturers, and distributors.

**Do you need a publicist?** Being represented by a publicist can make for an appealing presentation, but that ultimately makes no difference to the editors and writers of *Salon News*. If a salon owner and his or her salon are making news, doing creative work, or having experience the rest of the community should know about, we'll cover it. Letting us know that something happened or is happening is the most important thing.

**Some final pointers:**

◆ Even if you're not sure how newsworthy something is, get in touch with us. We're also here to help if you have a question or challenge but don't know where to turn for the information.

◆ Send us as much information—who, what, where, when, why and how—as possible. We frequently respond to letters and press releases with a telephone call, but it's helpful if all the necessary information is included.

◆ Please accompany all written submissions with a contact name (either the owner or the person responsible for the publicity), address, and daytime telephone number. Telephone calls/voice mail messages should refer to the specific topic and include the salon's name and a daytime telephone number.

◆ Be patient. We can't always respond to your queries immediately. That doesn't mean we won't.

*– Melissa Bedolis*
*Editor*

## *American Salon*

**What we are about:** *American Salon* (Fig. 7-8) is primarily staff written. However, readers are invited to submit photos and materials for two columns: "Hot Shots" and "FYI." Submitting materials for editorial consideration can be done as follows:

**PR opportunities:** For our monthly "Hot Shots" section, submit good quality color slides or black and white prints of contemporary

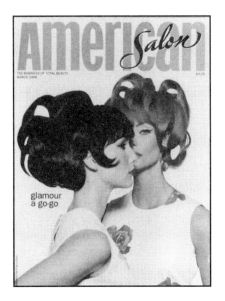

**Fig. 7-8** *If your salon is cutting edge, then* American Salon's *editors want to hear from you.*

**Fig. 7-9** *Every month, "Hot Shots" features photos submitted by* American Salon *readers.*

hairstyling (Fig 7-9). No Polaroid or photocopies are accepted. Be sure to label all of your materials, including each slide, with your salon name and phone number. Do not write directly on the back of a print. This damages the print and makes it difficult to reproduce in the magazine. Instead, write your salon name and phone number on a sticky label and apply the label to the back of the print.

Before submitting your work, look over the "Hot Shots" section of the magazine to get a feeling for the kinds of quality photography we strive to run. Make sure your materials fit the pattern you see. If you cannot imagine your work running among the shots printed in the magazine, the editors probably can't either. Generally, we look for an attractive model, good clean makeup, and original haircuts and styles with a finished look. Whenever possible, hire professional photographers and models. If that is not possible, use very young girls as models. Look for clear skin, symmetrical features, and a good smile.

*American Salon* does not pay for any submissions, but if we use your photo, we do credit you and send you a copy of the magazine (Fig 7-10). Please do not submit original work. Send duplicates. The reason for this is that we cannot acknowledge or return any artwork or submis-

**Fig. 7-10** *"Cutting Edge" features readers' hottest news.*

sions. We apologize for that, but there isn't time or manpower to keep up correspondence with every reader who submits materials.

Please do not call to check on the status of your submission. We use the work in the order it is received, generally, but are always on the lookout for particularly timely, trendy shots. We like hairstyles for men, women, children, older people, and people of every conceivable ethnicity.

Include details of your shoot, including credit information for the photographer, hairstylist, makeup artist, and information about your salon, including city and state and telephone number. Give us as much information as you can. Tell us what products you used, if relevant, and what you were trying to achieve and why.

The other section of the magazine that accepts reader submissions is "FYI," the salon and industry news section (Fig 7-11). We welcome salons to submit news and photos. Again, label everything. We especially like stories about expansions, charity events and involvement, new and unusual services, awards and citations, ideas for how to do things better in the salon, and important staff changes. We run news stories in the order they are received and edit them for timeliness and interest to

**Fig. 7-11** *"Salon of the Month," in* American Salon, *features salons with a unique story to tell.*

the general readership. Do not submit any photos that you need returned. We cannot acknowledge or return anything. If you do submit photos, please label them with your salon name and telephone number and provide, on a separate piece of paper, any identifications of individuals' pictures. Please make sure you have the correct spellings. We cannot call to verify spellings, so what you send us is what we run.

**Some final pointers:**

♦ Do not submit poems, show reviews, or any other unsolicited materials.

♦ Provide the answers to the following questions in every submission to any magazine: who, what, when, where, why, and how much.

♦ Label everything you send us.

*– Lorraine Korman*
*Editor in Chief*

110  Getting Known in Your Industry

**Fig. 7-12** SalonOvations *is the official publication of the National Cosmetology Association and particularly likes to feature work from younger or undiscovered talent. It's a great place to try your first pitch.*

## *SalonOvations*

**What we're about**: At *SalonOvations*, we are always looking for untapped talent of any age, and we especially like to encourage younger people to promote themselves and their talents (Fig. 7-12). We don't care if you don't have an international reputation. If your work is good, we'll consider it.

**How to work with us**: Know what we're doing. *SalonOvations* is always open to new ideas and we like to see what our readers are doing. To find out what we'll be working on, request an Editorial Calendar. Send a note requesting one and a self-addressed stamped envelope to *SalonOvations* magazine, P.O. Box 12519, Albany, NY 12212.

**What to send us**: We invite you to submit articles on relevant topics and pictures of work you're especially proud of. It must be well written in complete sentences that make sense. You can also be involved by dropping us a note to tell us your area of expertise and that you'd like to be a source.

We're always on the lookout for high-quality finished styles that can be blown up to a full page in the magazine. Often, we do use work to illustrate a story and much of that is submitted by readers.

**Dos and Don'ts:**

Do  take the time to put your slides in slide protectors.

Do  take the time to list on a separate sheet of paper all photo credits.

Do  take the time to photocopy your model releases and enclose a copy.

Do  take the time to put cardboard top and bottom to protect your precious slides from scratches.

Do  enclose the proper size envelope with the proper amount of postage if you'd like your work returned.

Do  be sure to include a telephone number where you can be reached during the day.

Don't  send work that is out of focus or very dark.

Don't  send handwritten notes unless your handwriting is legible.

Don't  ever send the only copy you have of your work—send a good duplicate and pay for quality.

*– Barbara Jewett*
*Managing Editor*

## *Nails* Magazine

**How to be involved:** Nail technicians can become involved in the magazine in a couple of ways: by providing or writing stores or by providing technical expertise by letting us know that you are available for interviews or doing the technical work for a photo shoot (Fig. 7-13).

**PR opportunities:** *Nails* is divided into specific sections:
- "Troubleshooter": This section is usually written in-house, but nail technicians can submit an idea on a common technical problem and how they deal with it. If artwork is available, please provide it, but we will usually shoot it ourselves if no photos are available.

**Fig. 7-13** Nails *magazine targets nail technicians in both nails-only and full-service salons.*

- "Letters": Readers can comment on particular issues or industry topics by writing directly to *Nails* magazine via mail, fax, or E-mail.
- "The Nails File": This is probably the best opportunity for nail technicians to get involved (Fig. 7-14). We accept all kinds of items for this extremely popular section, including personal anecdotes, poems, "Tip of the Month," salon events, customer service ideas, or retail success stories. Submit your ideas in writing.
- "Reader Nail Art": *Nails* publishes 6 to 8 reader nail art photos a month (Fig. 7-15). Submit quality color photos, a description of the art and the techniques involved, and your salon name and address. We have guidelines on how to best shoot nail art that you can get by calling our office at (310) 376-8788. It usually takes 4 to 5 months for material to be published and you will not be notified. If you want your photos returned, send a self-addressed stamped envelope with them.
- "Reader to Reader Advice": In this monthly column, a single question is posed and responded to each month, usually on business-related topics. You can respond via mail or fax.

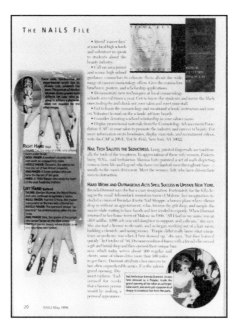

**Fig. 7-14** *"The Nails File" in* Nails *features news and tips from readers every month.*

- "Modern Nail Salon": This is a full-length feature on a single salon. If you're interested in being profiled, submit a short letter or essay on the salon's uniqueness, plus color photos to illustrate your points.
- "Cover Tech": We choose one reader each month to do the nails for our cover. Technicians must do exceptional quality work. Interested technicians should submit photos of their best work to me and follow up with a phone call. All photo shoots are done in California and *Nails* does not pay for travel expenses. (If you live out of state, you're responsible for paying for your own expenses.) However, we do pay the technician for doing the nails. The technician is also profiled in the "Cover Tech" section.

**How to get published:** We don't use many freelance articles, but we always welcome the opportunity to hear your ideas. Nail technicians should query in writing or call with specific article ideas. We are always available to talk to nail technicians and love to hear your ideas. It's always worth a call to make a suggestion or ask a question.

We are looking for information that nail technicians can use to build their businesses, save time, save money, or improve their skills.

**Fig. 7-15** *"Readers Nails Art" is one of* Nails' *most popular monthly features.*

Topics that are of interest include time-saving or troubleshooting techniques; nail art technicals and how-tos; salon and nail technician profiles (focus on what is different or what others can learn from you); customer service ideas; solving common salon problems (client related, employee related); health issues (although our health articles are usually written by experts in their fields, we like to hear from readers who have health or safety questions); and viewpoint or opinion pieces on industry issues.

– Cyndy Drummey<br>
Editor in Chief and Publisher

## *Passion and Coiffure Q*

**What we're about:** *Passion* and *Coiffure* Q are two of the most popular professional hairstyling magazines. Many salons around the world feature these two large format publications in their reception areas to use as style selectors for their clients. If you create high-quality hairstyles, then we would love to see them in *Passion* and *Coiffure* Q! If you're unfamiliar with hair fashion photography, then contact our office for a free submission kit.

**What to submit:** Photographs should be of professional quality, enclosed in protective packaging with full credits attached to each print or transparency. Photographs are acceptable in the following format only: 2" x 2.25" or 35 mm and black and white prints to a maximum of 8" x 10".

**How to get a good shot:** While the focus is on the hairstyle, the complete image, including model, makeup, background, and mood, should be of complementary interest. Priority will be given to work that is fresh, progressive, and commercial.

Hair fashions are constantly revolving and evolving. For that reason, we rarely make specific requests pertaining to styles. Creativity and commercial appeal, however, far outshine the ridiculously severe and avant garde. That's not to say that artistry is not encouraged—it is! But before you send in that photo of high twisted green curls, ask yourself a few questions: Would I wear this? Would my friends wear this? Would anyone wear this? If you answered "no way" to any of those questions, perhaps you might want to seriously think about toning it down.

When your style is ready to be photographed, double check for stray hair on clothing, face, and around the ears. Examine every angle to make sure that there are no unintentional gaps or partings or stains if the hair has been tinted. Also, look closely for visible grips and pins.

Clarity is the key when it comes to separating the good photos from the bad. Although hazy mood shots can be sultry and fun to use, you stand a much better chance of publication with a crisp, clear picture. A photograph that is properly focused and free of fussy backgrounds allows the viewer to savor the subject at hand, in this case, the hair.

Technically, the work must be nearly perfect. Even the tiniest nicks and scratches appear as major crevices when photographs are enlarged several times over. Find a photographer who will not only handle the finished product with care, but who specializes in hair and is not a newcomer to the publishing world.

**Submission deadlines are as follows:**

| | |
|---|---|
| Spring issue: | November 20 |
| Summer issue: | February 20 |
| Autumn issue: | May 20 |
| Winter issue: | August 20 |

**Fig. 7-16** *Working as an educator for a manufacturer can help you build your PR network. Oregon stylist Jack Sanders (left) from Steamboat Annie's and Chicago stylist Hildi MacCabe from Hair's Hildi! have both had their photo work and styling tips published in trade and consumer magazines as a result of their relationship with ABBA's products and assistance from the company's PR agency. They are shown here with Alan Benfield Bush, ABBA's president (center).*

If you survive the rigorous selection process, you can expect to be notified in a letter about 8 to 10 weeks after the deadline for which it was sent. If your photographs are not used, they will be returned by surface mail, meaning you may not see them for up to a year.

– Helen Moy
Editor

## Getting a Bigger PR Bang

Stylists who work with manufacturers full-time or even for one show a year can often parlay that into industry press (Fig. 7-16). Here's how:

**Fig. 7-17** *IBS American Team Y.E.S. (youth, energy, style) was formed to give young cosmetologists the opportunity to display their talents on the International Beauty Show's Fashion Theater and become educators and role models for others who are considering a career in cosmetology. Earn a place on the team and you're in for a year of PR, courtesy of IBS. The 1996 team included Allison G. Foster, Rob V. Horton III, Matt Jiovani, and Johnnie Morgan.*

First, editors are often invited to manufacturers' yearly educational events. If you're making a presentation or teaching a class, you can make yourself stand out in hopes of having coverage from the show or being tapped as a source for a feature at a later date. Often, if you call the manufacturer's publicist ahead of time, he or she will help you make the right connections.

Also, if you're not on stage but are attending, it's still a good opportunity to let an editor get to know your face. You're already showing a commitment to your profession by taking time to attend, which can get you one step ahead of your competitors (Fig. 7-17).

(Of course, all of these things can also be transformed into local PR.)

118  Getting Known in Your Industry

**Fig. 7-18** *Mark MacGregor, co-owner of Salon Enterprises' two chains of salons, has become so adept at PR that he was named chairman of the PR Committee for the International Chain Salon Association.*

## Join a Group, Get More PR

You can also give your industry PR program a boost if you're an active member of a professional association, such as Intercoiffure, the Salon Association, the National Cosmetology Association, the Nail Industry Association, etc.

First, joining and participating in these groups shows a commitment to your industry, which adds credibility to your PR case. Trade editors usually attend these groups' yearly meetings. If you're on stage or part of a presentation, chances are you'll make an impression.

Many of these associations offer their members the opportunity to participate in press-focused events from photo shoots to press conferences as a benefit of membership or a reward for participation (Fig. 7-18). This provides an opportunity to get help with your PR effort, usually at a much lower cost than if you did the same thing on your own. (Of course, the drawback is that you will be presented as part of a group, rather than as the main focal point of a press presentation.)

Plus, many magazines use the membership lists from these organizations to find new sources.

# Chapter 8

# Special Events, Charity Work, and Other Networking Opportunities

*T*he things that you do can be just as effective at generating PR as the actual press kits you put together. Two examples of this are special events and charity projects.

Special events are limited only by your imagination. It is important that your event has a real purpose, that you generate as much fanfare as possible, and that you invite as many people to witness and participate in the celebration as possible. Just some examples include:

- Holding a bridal fair at your salon during which brides and their attendants can view makeup techniques and hairstyles displayed with a variety of veils. Have the brides-to-be actually model the looks.
- Salon openings are always a grand occasion for a big event. Just make sure you're open and up and running for a month or two before putting on the big splash.
- Roll out a new service with plenty of fanfare. Again, have it up and running for at least a month so all the kinks are worked out before putting it on display.

**Fig. 8-1** The 135 employees of Tucson's four Gadabout Salons launched "Image Up," a program that provides 14 economically disadvantaged third and fourth graders with six years worth of new school clothes, hair cuts, and other gifts to help encourage positive self esteem and performance. The program is so revolutionary that First Lady Hillary Rodham Clinton sent "best wishes for a successful project Image Up" from the White House! Here, Gadabout CEO Pam McNair (center) welcomes kids to the Image Up program.

- Beauty parties are always popular. Let each staff member invite his or her 5 to 10 favorite clients, depending on how much space you have. Invite those clients to each bring a friend. Have them bring in their own styling tools and style their own hair as their hairdressers give tips. If your salon has nails, your nail technicians can "float" and do mini-manicures or paraffin treatments between styling sessions.

- Receptions to honor employees who have been promoted or reached a milestone in their employment can be big winners in many ways. First, rewarding employees for any reason helps increase staff morale. This celebration gives you a built-in reason to tell the press and your clients how great your staff members are.

- Salon anniversaries provide the perfect reason for a party. Why not let everyone know you've been in business for 10 great years? Hold one big bash or a month-long celebration with new surprises every day. Invite each staff member to plan the activities, including decor, prizes, and refreshments for one day or join with a few others to plan a full week. Make it great and your clients will celebrate with you!

# Getting Behind a Good Cause

Charity work, also known as cause marketing, involves raising money, collecting food or clothing, or calling attention to a cause that's important to a salon's staff and clients (Fig. 8-1). Options run the gamut from collecting food and money for an animal shelter in your own backyard to raising money for a sick child of a client or staff member to holding a major fashion show to raise money for juvenile diabetes.

The great thing about charity events is that you don't have to go it alone. There are always people willing to share the time and dollar commitments with you. Join forces with a women's group to put on a fashion show. Let them handle the fashions and models while you take care of the hair and makeup. This teamwork approach lets you do a bigger event and spread your good will, and good PR, across a wider audience.

On a more widespread level, most major manufacturers already have causes and charity programs in place. All you have to do is to tap into their expertise and bring their formats to your local level or join in with one of their national fund-raisers. In addition, distributors often organize fund-raisers to provide a format in which many of their customers can pool their resources for a major effort.

One great benefit from cause PR is that your clients can join in and help. Pulling together for something you both believe in helps strengthen your relationship with them and brings them closer to you. In addition, as more people join your cause, your chances of getting major press increase, especially when your clientele includes local media.

# For Richard Calsacola, Good Deeds Turn Into Good Press

Chances are, you've heard of Richard Calcasola. That's because the owner of Maximus Salon & Spa in Merrick, NY, is a firm believer in giving back to the community. He achieves this through Maximus' participation in local charity-related activities, providing informative workshops and seminars, or creating his own fund-raising events. These events then naturally turn into publicity, with a little work.

To raise funds for a local community nursery school, Richard created a "Maximus Evening of Total Beauty," which allowed individuals to make a single contribution in exchange for as many salon and spa services as they could fit into the 3-hour event. Richard donated his facility and products. Thirty-six members of the staff enthusiastically volunteered their time to perform services, and hundreds enjoyed Maximus' fine services. The results? Several thousand dollars were raised for a worthy cause. In exchange for doing a good thing for the community, Maximus received pre-event coverage in local papers and on local cable and radio stations.

After the event, a release was written about the successful fund-raiser, sent out to the media, and picked up by both the local press and trade publications.

Maximus' involvement with any fund-raiser or community-minded event lends itself to press coverage through the proper execution of effective PR, coordinated and generated through their PR consultant, Victoria Spedale, principal of Victoria Communications in West Islip, NY.

The following gives a brief list of other creative Maximus-sponsored events that have generated successful PR results:

- **Silent Art Auction:** Maximus was converted into a gallery, featuring paintings, mobiles, sculptures, jewelry, and other works created by 65 staff members. Funds were raised for four local charities, including juvenile diabetes.
- **Cut- & color-a-thon:** Prior to participating with other salons in the tri-state HairCares annual fund-raising event, Maximus hosts this event to raise additional monies for this charity. Staff members donate their services for customers to receive a $15 haircut or $25 coloring during a designated day.
- **Free health workshops:** Each month, the spa hosts a specialist to enlighten a particular segment of the community on a specific topic, for example "Nutrition for the Woman

over 50," "Weight Loss in the Right Places" or "Skin and Body Care for the Pre-Teen."

- ◆ **Bridal beauty expo & champagne party:** Each June, Maximus gives preregistered brides and their bridal parties a free pre-event hair consultation and trial makeup application. Local bridal-related sponsors donate raffles and door prizes.
- ◆ **Back to school fashion show:** Before school begins, Maximus seeks local teens to model the latest in makeup, hairstyles, and fashions (provided by local merchants) for the coming high school/college year. For many teens, this is their first introduction to salon and spa services.

## Grass Roots Awareness: Robert Lamorte

For Chicago salon owner Robert LaMorte (Fig. 8-2), joining forces with Y-Me was a natural move more than a decade ago. This grass roots breast cancer awareness organization, which is based in Chicago, holds an annual fashion show fund-raiser. The president, who was also a long-time client, asked Robert to do the hair, and he immediately said, "yes." Knowing that the majority of his clients and staff at Robert Jeffrey Hair Studios are women, he thought it was the perfect step to turn his energy toward helping an organization that supports women's interests.

Although Robert's relationship began with styling the hair during this lunch, it eventually evolved into a major fashion show that Robert hosted and invited other leading salon owners from Chicago Intercoiffure to join him. As these hairdressers stepped out from behind the curtain, they presented their artistry to more than 700 leading Chicago consumers and raised money for a good cause.

**Fig. 8-2** Robert LaMorte (back row, second from left) and Intercoiffure Chicago

What began as, and still is, a labor of love for this salon owner, also turned into a national PR coup that's still running strong years later. What's even better: Since Robert introduced Y-Me to his industry, everyone from local salons to *American Salon* magazine has jumped on the Y-Me bandwagon. Following Robert's trail, they've continued to raise the money for this worthy cause. And, they invited Robert as an honorary guest at their first event, called the Hair Do, in New York City during the International Beauty Show.

As a result, what began more than a decade ago with a salon owner doing hair behind the curtain for a local fund-raiser, has turned into a national event that's supported by an entire industry and appreciated by women everywhere.

# Sample Special Event Press Release

**FOR IMMEDIATE RELEASE**
Contact: Robert LaMorte (Robert Jeffrey Hair Studio)
     (123) 456-7890

**INTERCOIFFURE CHICAGO PERFORMS THE ULTIMATE CLIENT SERVICE, RAISING $40,000 FOR BREAST CANCER AWARENESS**

CHICAGO – DATE–Six Chicago salon owners and their staffs closed out National Breast Cancer Awareness month by performing what might be the ultimate in client service, raising almost $40,000 for breast cancer awareness.

All proceeds from their "Hair Fashion" stage presentation were donated to Y-Me, the largest national, nonprofit, grass roots organization that promotes breast cancer awareness and offers support to women with this disease. The salon owners are all members of Intercoiffure, an international organization of the most prestigious hairdressers in the world.

The Chicago members' gala benefit, performed for an audience of clients, friends and fellow Intercoiffure members kicked off a nationwide salon industry campaign to raise money for breast cancer awareness.

Salon owners Jerry Gordon, J. Gordon Designs; Tom Hayden, Mr. T. and Company; Robert LaMorte, Robert Jeffrey Hair Studios; Warren Michaels and Tom Ramagnano, Michael Thomas Salons and Day Spas; and Mario Tricoci, Mario Tricoci Salons and Day Spas; and their staffs showcased their own commercial and artistic hair fashion statements during runway presentations. To show his support and enthusiasm for the program, special guest Intercoiffure President John Jay served as co-master of ceremonies with Y-Me President Pam DeLuca, and also did hair for the grand finale, which featured special guests Ballet Chicago. Jay also encouraged salons across the country to follow the efforts of the Chicago group.

Ann Marcou, co-founder of Y-Me and a client of LaMorte, introduced both shows.

"We did this for our clients to say thank you very much and to give back to them for their support throughout the years," said the members of Intercoiffure. "We wanted to share with the Chicago community – particularly women – how much we as an industry care about them and to help and support them, in addition to thanking them for their business. Collectively, our efforts in Chicago will be an example to salons in the rest of the country, who will organize to raise money for women's issues."

# Chapter 9

# Where Do I Go From Here?

## Your Image Starts and Stops Here

Now imagine that it's 6 months down the road. You've worked very hard. You have a successful PR program up and running (Fig. 9-1). Your new quick service perm was mentioned in your local newspaper's fashion pages; you participated in the Mirror Image Breast Cancer Awareness program and your local radio station did a remote broadcast from your big event. In addition, *Modern Salon* featured you in a follow-up feature to this program. You had a photo from your first shoot published in *American Salon*'s Hot Shots, as well as in *HairDo* and *Celebrity Hairstyles*. You traveled to the International Beauty Show in New York City and brought back a trend report. Your local gossip editor ran two lines about it, which generated more than 50 calls! Finally, *Salon News* interviewed you for its article on, "How I Started a New PR Program For My Salon."

Eight placements in 6 months. That's an excellent response. Congratulations!

So now what do you do? Brag about it to your clients! Get an extra bang from your PR successes by promoting them in your salon. The benefits are threefold:

- ◆ Your clients will be impressed that the editors like you! (They don't know how hard you worked to make these placements happen!) They'll probably even brag about you to their friends, which means an increase in personal referrals.

**Fig. 9-1** *When you do anything special, make sure you tell your clients, as well as the press. Here, stylist Edwin Fontanez works on stage for Matrix at the International Beauty Show. In addition to receiving press for his work, he spent the next month impressing his clients through his internal PR. Photo: Mike Kentz.*

- Your clients will be able to read good things about you as they sit in your salon. These positive messages will reinforce their belief that you're the best! And people love to be associated with winners. Again, they'll tell all of their friends.
- It will help generate pride among your staff. When they work all day surrounded by press that says you and your business are terrific, they'll be proud to work in such a winning environment.
  Here's what to do:
- Have your actual placements framed. Enlarge them if necessary and make sure that the name of the publication they were featured in is visible. Local framing shops or copy centers can help you.
- If you have a salon newsletter, run a regular column that quotes all of your PR placements. For example, "*The Daily City Paper* says that, 'XYZ salon is …' And, we're so proud to tell you that our

newest trend photos were pictured in *American Salon* magazine." Then, include a picture of the photo that ran in the magazine. Chances are, as your credentials grown, you'll see some clients who haven't been in for a while come back just to see what's going on.

- If you don't have a newsletter, after you have several mentions, you can inexpensively put a flyer together that plays up your press. Then, pass it out to your staff and clients. Post them in your changing area. Even mail them to clients you haven't see for a while and your local media. Again, chances are you'll see them during the next few months just so they can check you out.

- Once your program is really going, you might even want to have a client contest that works as follows: The first client who sees a new placement or hears you on the radio and calls you wins a prize from your salon—maybe a free haircut for him or her and a friend or a prize package of various add-on services. Make this contest ongoing. It really gets your clients excited and involved with you. Then, you can even publicize your contest and your name of winners!

Beyond that, use your creativity! Just get the message out!

# PR Ideas From A to Z

After working your way through this book, you're now armed with the knowledge of how to publicize your salon and its activities. Now, it's time to do it. To get you started, we've included one more list of ideas that is press worthy. Use them as is or let them kindle your imagination and come up with your own. You might even want to hold a staff contest to see who can come up with the most original PR idea.

Many of your regular activities can be big PR getters, when you tell someone about them. Here's a list of things you can do. Then, tell someone about them. Chances are, you'll see your name in all the right places in no time!

- Adopt a women's shelter, nursing home, child care center, or animal shelter. Get your clients and friends involved. You can visit your adoptees, raise money, create awareness...whatever is needed most.

- Attend a trade show. Then, let your local fashion media know what you learned and how it applies to your market (Fig. 9-2).

- Also, let your local gossip/society editors know where you went and what important events you attended.

**Fig. 9-2** *Candy Shaw (left) of Jamison Shaw Hairdressers in Atlanta and Jesse Briggs (right) of Yellow Strawberry Global Salons in Ft. Lauderdale not only attended the International Beauty Show in New York City, they worked it for BES Regal Haircolor. As a result, they were able to network with beauty editors and fellow professionals simultaneously.*

- Add a new service to your salon menu. Tell your local media how this new service represents a trend in relaxation services, hair color, or whatever. Play up the service and invite your local media in to try it.
- Award a scholarship. It can be to beauty school, to college, to a trade school—whatever works best in your area. You don't have to pick up the entire tuition. Anything from a $50 savings bond to a $500 check would be appreciated. No matter what the amount, the fact that you're doing something really puts you in a good light. Pick press-friendly criteria and send out pictures of yourself and the winner. You might want to have a party at your salon during which you make the presentation and invite the media.
- Be as unique and creative as possible. One of the most famous salon industry PR moves of all time was when the late Robin Weir, hairdresser to Nancy Reagan and other famous politicos, had a pedicure done on an elephant. The photo ran from coast to coast.
- Beat the weather. If it's been extremely cold, hot, rainy, or foggy for weeks with no end in sight, use it to your advantage. Hold a

party, have a countdown, run a special or promotion, and tie it in to the situation. You can even call your local TV weather departments. Many, especially in the morning, do impromptu on-site broadcasts with entertainment value.

- Bond with teens. Be the first one to congratulate the new class officers, new cheerleaders, homecoming court and Valentine's princesses with offers of your service.

- Bring on the brides! Network with a local bridal fair or create your own "Beauty Fair" for brides and their attendants in your salon. Offer free makeup and hairstyling consultations for the bride if she books consultations for the bridal party. You can even make your salon available for bridal showers. Of course, issue a press release to describe the fun events you're offering. If you're doing a big bridal fair, call the local radio station and pitch them on doing a remote broadcast from your fair.

- Buy a new piece of equipment that is seldom seen in your town (of course, only if it's appropriate for and in line with your business plan). New space age perm and color processors, deluxe pedicure thrones, and skin scanners for your esthetics room are some examples. Then, tell your local media how "space age technology" or "deluxe pampering" is the newest trend, emphasizing how this will help you give your clients what they want—and what they can't get anywhere else. Invite editors/reporters in to try it for themselves.

- Call in to radio or TV shows that are addressing beauty or fashion topics to answer a question or offer your expertise.

- Celebrate something off the wall that's keeping with your image and sure to attract attention. Select an event that will get a lot of media attention, but that most salons wouldn't think of. Elvis' birthday, the spring or fall fashion collections, Ground Hog Day, Super Bowl Sunday are some ideas. Give the normal revelers a new kind of place to celebrate.

- Celebrate an anniversary and invite the whole town! Whether it's your tenth year in business or the first year anniversary since you added day spa services, it pays to let everyone know. Create special in-salon events and promotions and invite the local media. Send a press release to national media, too. Just give it an interesting twist. For example, if it's your tenth year in business, are there any trends that are back again that were hot when you first opened? Play them up!

- Collect photos of famous hair. Cover the faces and have clients try to guess the person by the hair. Offer a prize per day for a week or month.

- ◆ Color your world. Hold a color-a-thon to raise money for your favorite charity, while calling attention to how hair color can beautify someone's world. Challenge all of your clients and everyone in town to stop by and try-out hair color. Any color, from a temporary wash-out spray to an all-over bright glazing to just six foils around the face, goes. Encourage clients of all ages, particularly men and teens, who might try it for a good cause and then never give it up again. Provided you and your staff can get enough guarantees to expect a good turnout, announce to the media that you're attempting to break your city's record for consecutive hours coloring hair. (Chances are there's nothing recorded so it's yours for the effort. Double check just in case.) Set a goal—300 people, 24 hours—and award a prize for the person who puts you over your goal. Have your makeup staff ready with quick mini-makeovers to show them how color can brighten their entire outlook.
- ◆ Comment on current events. If it's getting close to election time, find trends in the candidates' hair. Do you notice a pattern in hairstyle, color, length, or level of hair loss for winners or losers?
- ◆ Complement 25 people you don't know who are not clients about how they look. Do it only for legitimate reasons. Spreading good cheer and positive feelings will bounce back to you in the form of good PR.
- ◆ Create a community involvement suggestion box. Ask staff and clients to drop in suggestions of charities and events they'd like to see you support.
- ◆ Cut a local or national celebrity's hair. To protect the relationship, make sure you get permission before you publicize who's hair you do.
- ◆ Display interesting, press-worthy materials that relate to your business or clients. For example, if you do the hair for local TV anchors, get them to autograph their press photos. Frame them and create a central display.
- ◆ Do a photo shoot.
- ◆ Donate gift certificates for services at your salon to a local charity auction that supports causes you and your clients support. Make sure you receive credit in the program.
- ◆ Educate your staff about a new service or trend. Then, publicize the fact that you did the education. For example, invite your hair color supplier in to do a hands-on class on the newest hair color techniques. Invite popular students who are the fashion leaders at their schools to be hair color models. Then, take photos of them.

Send them to your local paper, as well as their school papers. You might even want to invite your local fashion editor or TV reporter to attend to talk about the trend and the excitement your salon is creating.

- Enter the North American Hair Styling Awards and become a finalist. Deadline for entry is usually in the early fall. For rules, guidelines and helpful hints, call 1-800-468-BBSI

- Form a fashion advisory council. Invite the local fashion editors, local fashion and jewelry designers, photographers, makeup artists, fashion stylists, models, top department store fashion stylists, and representatives of any locally based beauty companies to participate. Have a purpose beyond networking: to organize a benefit, to put out a quarterly city-wide head-to-toe fashion advisory, or to define and answer the area's top 10 fashion questions. Even if you can just get five people together to start, spread the word about what you're doing and your group will grow over time.

- Give away a prize a day for an entire month to celebrate a specific occasion. It can relate to your salon, your city, a local sports team's victory, or even the end of winter. Hint: Do it at a time of the year that is usually slower than normal. You can generate press and new business simultaneously.

- Give out your business card in a friendly way wherever you go. Encourage your staff to do the same.

- Give your business card to any well-connected clients and ask them to pass your cards along when appropriate.

- Go online. Become a source of beauty information on the InterNet and then promote your computer-savvy salon.

- Hold a contest among your staff to generate PR ideas. Even if it's not on target, you can fine tune it and their enthusiasm will contribute to your program's overall success.

- Hold a cut-a-thon.

- Hold a "Teen Night." Invite local teens to your salon to learn about beauty and personal care. Make sure to invite your local beauty/fashion writers, as well as reporters for the students' school newspaper and yearbook.

- Invent a product that solves a problem your clients are having or that makes it easy for them to jump on top of a trend with little investment. Think of how the life has changed for the hairdresser who invented Topsy Tail! If you want to keep it in your home town, go for something simpler, using common household items to set or hold hair or as a cute accessory.

- Invite beauty school students to tour your salon for a career day.
- Join a national professional organization. Many salon professionals get national press in conjunction with their connections with the National Cosmetology Association, Intercoiffure, the Salon Association, and others. Many of these organization do photo shoots that you can take part in for a small fee or as part of your membership. They also hold national and sometimes even international meetings. When you attend these, you can publicize what you learned to your local press.
- Join the Cosmetology Advancement Foundation's Role Model program. If you're selected as a role model, CAF's publicist will set up interviews for you to talk about the rewards of a career in cosmetology.
- Key in on the most common beauty/grooming problem or pet peeve of groups of people, including women, men, teens, or children. Do a survey or use your own instinct. Announce the top pet peeves then give your solutions to what they can do about them. Make your answers informational and education, but above all, make them fun.
- Let your clients know that you have launched a PR program. It pays to find out who they know, where they and their families work, and how they can help you.
- Link up with a manufacturer. Many offer you assistance with PR on a local or national level when you attend their educational events, assist their design team, and even use their products. Let your distributor know you are interested in any opportunities that are available and be persistent.
- Link up with a local fashion designer, health and fitness expert, dermatologist, nutritionist, or any other professional whose information would be useful to your clients and is complementary to your image.
- Make over anyone. Everyone who's willing and motivated is fair game: a bride to be, a new mom, a graying executive, a whole football or volleyball team or cheerleading squad, a zoo animal, an entire family, sets of twins, whoever and whatever fits your target clientele and is appropriate for the season. Try sweethearts for Valentine's Day or teens for back-to-school. Make over police officers for crime prevention week. The press loves makeovers. Make them interesting, easy, fun, and commercial.
- Mentor a beauty school student. Report on the process throughout. When he or she graduates, throw a party at your salon to welcome him or her to your profession.

- Name new services with a particular mind to PR. Catchy names for services call more attention to the service than boring, straightforward ones do. For example, would you rather have "color glazing" or "single process hair color"?
- Offer your services to your local speaker's bureau. Provide a list of beauty, fashion, and personal care topics you can speak about that are in line with your image. Chances are, if you're booked for an event, the organizers will publicize it for you. Check in advance and if they don't, handle the PR yourself. Also, make sure you know who will be in the audience. If local writers, news anchors or DJs are in the group, make sure you find out who they are and introduce yourself.
- Open your doors. Make your salon available to use for meetings or private parties.
- Participate in *Modern Salon*'s Mirror Image program and turn your salon into a breast cancer awareness center. Hold an evening get together or keep the program going daily for a month. Think of the educational benefits you can provide for your clients and staff. The rewards of good press often come to those who care.
- Promote a Pets and Pedicures day, if you're a dog or cat lover. Team up with a local animal groomer and offer a two-for-one special. Clients get a pedicure while their dog or cat (limit it to small pets) gets a nail trim. Have someone take snapshots of the well-groomed duos to send to the people with an invitation to come back to the salon, as well as to send to the press.
- Pick a theme for your salon. Play it up to the hilt, then publicize it.
- Poll your clients or staff about their favorite 10 celebrity hairdos. Then, hold a look-a-like contest to find people who look like your top 10 stars. Do a makeover on each one or show how they can change their looks easily with no commitment. Post photos of the winners along with those of the celebrities they look like in your salon. Of course, you'll announce your contest to the press ahead of time and then call everyone once you have your winners.
- Promote a staff member. Then tell the world that staff member's story, what he or she specializes in and why he or she is special. What an extra reward and boost of support for the staff member, too!
- Put on a fashion show at a local mall. Team up with a clothing store that's in sync with your clients' style and stage a production to show off the latest fashions for back to school, Mother's Day, or any holiday. Get some of your clients to model!

**Figs. 9-3, 9-4** *To appeal to younger, health conscious clientele, Jesse and Flo Briggs' Yellow Strawberry Global Salons sponsor young athletes, including Kelly Moore, the world's top ranked wind surfer, and Mary Ellen Clark (below with Jesse), the dominant force in women's diving today. Kelly sports the YSGS logo on her sail (right), while Mary Ellen wears it on her swimsuit. Every time they are featured on TV, in the newspaper, or in magazines, Yellow Strawberry's logo is seen by millions.*

- Query a trade magazine editor with an idea for an article. Talk about how you solved a common problem in a way that's different from what everyone else does. A retail promotion, training program, staff reward system, or client incentive program are all good bets if you achieved great results.
- Raise money for a local homeless shelter or any other charity you support. Invite your clients to help and let your local media know what you are doing and how people can be involved.
- Remodel your salon for a reason. Then tell the world why and what benefits this brings to your clients and staff.
- Research the history of beauty and fashion in your town or city. Then, go on a tour of local media, schools, women's groups, and the like to talk about what you've learned and what it means. You can use models to demonstrate what used to be. Then, show where it's going.
- Set a trend. If it fits your salon's image, be the first to go red, the first to have pastel nail polishes, or the first to wear a new bouffant. People who look interesting, without going overboard, tend to be thought of as interesting—a key to getting good press.
- Solve a problem. Any problem. Link it to the season or to a particular problem many of your clients are having. For example, once the pools open, send out a press release for ways to remove chlorine buildup from the hair. Does your town have hard water or water with minerals? Tell how that can affect hairstyling, color, and perms, and give solutions.
- Sponsor a local athlete (Figs 9-3, 9-4). Do you have an aspiring competitor in your area? It could be in any field; ice skating, gymnastics, wind surfing, roller skating, even bowling, whatever's popular in your part of the country. By contributing a small amount of money to help that person train or sponsoring a single event, many will agree to wear your logo during competition. Think of the exposure you could get while helping people achieve their goals! And, of course, you can announce your sponsorship to a variety of media.
- Start a newsletter that tells what you're doing. Even if it's on just one piece of paper, it will help create excitement about what you're doing.
- Style a local or national celebrity's hair.
- Style hair for a charity fashion event.
- Survey your staff. Find out the top 10 questions their clients ask them. Then publicize the results.

- Take an editor to lunch.
- Take a poll. Ask your clients and staff what local or national celebrities they think have the best hair. Publicize your results.
- Teach a class for other beauty pros, for your clients, or any group who would like to attend. You can talk about anything from how to do the newest styles to how to match hair with fashion. In fact, you can even schedule a whole curriculum so people know what topics they can look forward to.
- Travel abroad or to New York or Los Angeles for vacation or a trade show. When you get back, let your local beauty, fashion, and gossip writers know where you went and what you learned. Be their eyes and ears for fashion and beauty trends.
- Unmask a myth. Many people still believe that brushing their hair 100 strokes a night is healthy. You know that's not necessarily true. Show them why and offer alternative grooming practices. Does your town have an older population who thinks they are doomed to thinning, graying hair? Show them how to transform their tresses back to the hair of their youth. If teens in your area think they will get acne no matter what, have your esthetician do a class to show them how to manage it. No matter what myth you choose to debunk, the information you provide will work nicely into a press release.
- Visit a local high school to talk about the variety of careers students can pursue with a cosmetology license. Bring your staff members to talk about working in a salon, on photo shoots, on movie sets, for a manufacturer on stage or in the classroom, and as a salon owner, manager, or merchandiser. You'll help spread valuable information while generating great PR for your salon and your profession. You can also write to the Cosmetology Advancement Foundation for brochures that explain career opportunities. The cost is $20 for 100 brochures. You can write to CAF at 208 East 51st Street, Suite 143, New York, NY 10022.
- Volunteer to do makeovers at a home for the elderly, hospice, or children's hospital.
- Watch an awards show and then comment on the hairstyles. For example, the day after the Academy Awards, have your list ready to go of the best hair, best total image, worst hair, best updo, best haircolor, or men's look of the night, etc. If your salon is more hip or youth oriented, try the same idea for the MTV Awards. Here, the key is timeliness. Do the legwork ahead of time. Get a list of nominees and presenters and pull out the "usual suspects."

- Write a thank you note or send flowers every time you appear in your local media.
- X-pand your knowledge of what the press wants. Invite them to your salon or to lunch and focus the entire conversation on them. For that one time, don't talk about your salon at all unless you're answering a question.
- X-pand your media list. Visit a larger newsstand or bookstore and investigate your local cable systems to find three or four new media outlets that serve your target audience.
- Y-Me. Join forces with this national breast cancer educational organization. Volunteer to pass out literature in your salon. Do the hair for an annual fund-raiser. Invite members into your salon to help them look and feel their best.
- Zero in on one thing that makes your salon completely unique. Play it up to the hilt.

## Appendix 1

# Local Media List

Use this form to set up your own local media list, including newspapers, radio programs (there may be several programs at one station), television programs and news segments, and even any local organization's newsletters that accept PR materials.

Be sure to include important information you learn about each contact as you make them, such as a particular day of the week or time they do or don't want to be contacted. Also, write in any regular beauty/fashion sections. For example for your local newspaper, Fashion Section every Thursday. Or, for your local morning TV program, makeover segment on the third Tuesday of each month.

| PUBLICATION | CONTACT | MY NOTES |
| ADDRESS | NAME/ | |
| PHONE/FAX | TITLE | |

*Local Media List (continued)*

| PUBLICATION<br>ADDRESS<br>PHONE/FAX | CONTACT<br>NAME/<br>TITLE | MY NOTES |

*Local Media List (continued)*

| PUBLICATION ADDRESS PHONE/FAX | CONTACT NAME/ TITLE | MY NOTES |
|---|---|---|
| | | |

## Local Media List (continued)

| PUBLICATION<br>ADDRESS<br>PHONE/FAX | CONTACT<br>NAME/<br>TITLE | MY NOTES |
|---|---|---|

*Local Media List (continued)*

| PUBLICATION<br>ADDRESS<br>PHONE/FAX | CONTACT<br>NAME/<br>TITLE | MY NOTES |
|---|---|---|
|  |  |  |

## Local Media List (continued)

| PUBLICATION ADDRESS PHONE/FAX | CONTACT NAME/ TITLE | MY NOTES |
| --- | --- | --- |

# Appendix 2

# Popular Consumer Magazines

Following are some of the most popular national consumer magazines. This list contains major titles and is representative of the types of magazines you might decide to target based on your salon's image.

Keep in mind that often the type of information you send to a beauty or fashion magazine will be different from the tips sent to a service magazine. Target the teen magazines with information for 12 to 19 year olds if this group is included in your clientele. There are also quite a few magazines for the African American market, as well as new titles for the Hispanic/Latino and Asian markets almost monthly. If you service those groups, it pays to research what's new. Start by asking your clients what magazines they like to read.

Many popular consumer magazines are based in New York City. For salons on the West Coast, many of these same publications have editors on the West Coast, so it might pay to ask an editor if there's a contact you can meet with in your area. Although they don't handle beauty exclusively, many market editors research several topics in their markets and report back to their main offices.

Once you become a source for a magazine, you will likely add several other contacts to your list. Often, freelance writers will handle research and writing, so you'll want to add them as you get to know them, too.

As always, read magazines that are on your "hit list" every month and check the masthead or staff listing to make sure you know about the newest contacts.

| PUBLICATION ADDRESS PHONE/FAX | CONTACT NAME/ TITLE | MY NOTES |
|---|---|---|
| *All About You Teen* 6420 Wilshire Blvd. Los Angeles, CA 90048 phone: (213) 782-2950 | Roxanne Cameron Editor in Chief | Teens |
| *Allure* 360 Madison Avenue New York, NY 10017 phone: (212) 880-5595 | Martha McCully Beauty Editor | Beauty/Fashion |
| *Black Elegance* 475 Park Avenue South New York, NY 10016 phone: (212) 689-2830 | Sonia Allyne Editor | African American |
| *Bridal Guide* 3 East 54th Street, 15th Floor New York, NY 10022-3108 phone: (212) 838-7733 | Jacklyn Monk Beauty Editor | Bridal |
| *Brides* 140 East 45th Street New York, NY 10017 phone: (212) 880-8331 | Denise O'Donaghue Beauty/Fashion Editor | Bridal |
| *Cosmopolitan* 224 West 57th Street New York, NY 10019 phone: (212) 649-2000 | Andrea Pomerantz Beauty Editor | Fashion/Beauty Lifestyle |
| *ELLE* 1633 Broadway New York, NY 10019 phone: (212) 767-5800 | Jean Godfrey June Beauty/Fitness Director | Fashion/Beauty/ Lifestyle |
| *Essence* 1500 Broadway New York, NY 10690 phone: (212) 642-0600 | Jenyne Raines Beauty Associate | African-American |

Appendix Two **151**

| PUBLICATION<br>ADDRESS<br>PHONE/FAX | CONTACT<br>NAME/<br>TITLE | MY NOTES |
|---|---|---|
| *Face*<br>31727 Pacific Coast Highway<br>Malibu, CA 90265<br>phone: (310) 589-2600<br>fax: (310) 457-0535 | Shi Kagy<br>Editor | Asian |
| *Family Circle*<br>110 Fifth Avenue<br>New York, NY 10011<br>phone: (212) 463-1000 | Linda Moran Evans<br>Beauty Editor | Women's Service |
| *First for Women*<br>270 Sylvan Avenue<br>Englewood Cliffs, NJ 07632<br>phone: (201) 869-6699 | Barbara Brown<br>Beauty Editor | Women's Service |
| *Glamour*<br>350 Madison Avenue<br>New York, NY 10017<br>phone: (212) 880-8211 | Leslie Seymour<br>Beauty Editor | Beauty |
| *Good Housekeeping*<br>959 Eighth Avenue<br>Suite 641<br>New York, NY 10019<br>phone: (212) 649-2412 | Karyn Repinski<br>Beauty Editor | Women's Service |
| *Harper's Bazaar*<br>1700 Broadway<br>New York, NY 10019<br>phone: (212) 903-5000 | Annemarie Iverson<br>Beauty and Health Editor | Fashion |
| *Ladies Home Journal*<br>100 Park Avenue<br>New York, NY 10017<br>phone: (212) 953-7070 | Lois Joy Johnson<br>Beauty & Fashion Director | Women's Service |
| *Mademoiselle*<br>350 Madison Avenue<br>New York, NY 10017<br>phone: (212) 880-8569 | Jane Larkworthy<br>Associate Beauty Editor | Beauty/Fashion |

| PUBLICATION<br>ADDRESS<br>PHONE/FAX | CONTACT<br>NAME/<br>TITLE | MY NOTES |
|---|---|---|
| *Marie Claire*<br>250 W. 55th Street<br>5th Floor<br>New York, NY 10019<br>phone: (212) 649-4450 | Alexandra Parnass<br>Beauty & Fitness Director | Fashion/Beauty |
| *McCall's*<br>110 Fifth Avenue<br>New York, NY 10011<br>phone: (212) 463-1435 | Colleen Sullivan<br>Beauty Editor | Women's Service |
| *Mirabella*<br>1633 Broadway<br>New York, NY 10019<br>phone: (212) 767-5817 | Mary Lisa Gavenas<br>Beauty Director | Women's Issues |
| *Modern Bride*<br>249 West 17th Street<br>New York, NY 10011<br>phone: (212) 337-7113 | Martine Niddam<br>Beauty Director | Bridal |
| *New Woman*<br>215 Lexington Avenue<br>New York, NY 10016<br>phone: (212) 251-1500 | Clare Cannon<br>Beauty Editor | Women's Issues |
| *Redbook*<br>224 West 57th Street<br>New York, NY 10019<br>phone: (212) 649-3432 | Cheryl Kramer<br>Associate Beauty Editor | Women's Service |
| *Self*<br>350 Madison Avenue<br>New York, NY 10017<br>phone: (212) 880-8800 | Janet Carlson Freed<br>Beauty Director | Lifestyle/Beauty |
| *Seventeen*<br>850 Third Avenue<br>New York, NY 10022<br>phone: (212) 407-9700 | Elizabeth Brous<br>Beauty Editor | Teen |

| PUBLICATION<br>ADDRESS<br>PHONE/FAX | CONTACT<br>NAME/<br>TITLE | MY NOTES |
|---|---|---|
| *Teen*<br>437 Madison Avenue<br>New York, NY 10022<br>phone: (212) 935-9150 | Mary Rose Almasi<br>Beauty/Fashion Editor | Teen |
| *Vogue*<br>350 Madison Avenue<br>New York, NY 10017<br>phone: (212) 880-8818 | Amy Taran Astley<br>Beauty Director | Fashion |
| *W*<br>7 W. 34th Street<br>New York, NY 10001<br>phone: (212) 630-3557 | Dana Wood<br>Beauty Editor | Fashion |
| *Woman's Day*<br>1633 Broadway<br>New York, NY 10019<br>phone: (212) 767-6000 | Rebecca Nelson<br>Beauty/Fashion Editor | Service |
| *Woman's World*<br>270 Sylvan Avenue<br>Englewood Cliffs, NJ 07632<br>phone: (201) 569-0006 | Elaine Karten<br>Associate Beauty Editor | Service |
| *YM*<br>685 Third Avenue<br>New York, NY 10017<br>phone: (212) 878-8700 | Cara Kagan<br>Beauty Editor | Teen |

| PUBLICATION<br>ADDRESS<br>PHONE/FAX | CONTACT<br>NAME/<br>TITLE | MY NOTES |
| --- | --- | --- |
| | | |

# Appendix 3

# Consumer Hairstyling Publications

Consumer hairstyling publications are found in grocery stores, drug stores, mass merchandisers, on newsstands, and in a variety of other outlets. Many salons even keep some of these publications in their reception areas so their clients can use them as style selectors.

These hairstyling publications are wonderful outlets for your photos. They use many photos each issue, so as long as you meet their needs, they're likely to use your work again and again. They love to develop close relationships with hairstylists across the country, so chance are, they'll be open to your calls. These companies publish many different titles, several times a year. They reach tens of millions of consumers, which means you can become "a name" quickly.

| PUBLICATION<br>ADDRESS<br>PHONE/FAX | CONTACT<br>NAME/<br>TITLE | MY NOTES |
| --- | --- | --- |
| Harris Publications Inc.<br>1115 Broadway<br>New York, NY 10010<br>phone: (212) 807-7100 | Mary Greenberg<br>Editor | |

Titles include *Celebrity Hairstyles, HairDo Ideas, New Ideas for Hairstyling, Short Hair Styles, Soap Star Hair Styles,* and more.

| PUBLICATION<br>ADDRESS<br>PHONE/FAX | CONTACT<br>NAME/<br>TITLE | MY NOTES |
|---|---|---|
| Harris Publications, Inc.<br>1115 Broadway<br>New York, NY 10010<br>phone: (212) 807-7100 | Anne Charles<br>Editor | African American<br>titles |

Titles include *Black Hair Care* and *Black-Tress*

| | | |
|---|---|---|
| GCR Publishing<br>1700 Broadway<br>New York, NY 10019<br>phone: (212) 541-7100 | Sandra Kosherick<br>Managing Editor | |

Titles include *Black Hairstyles, Complete Short Hair Styling Guide, Complete Hair and Beauty Guide, Instant Hair Styles, Short Hair Styles, Step-By-Step Hairdo Ideas*, and more.

| | | |
|---|---|---|
| Sophisticate's Hairstyle Guides<br>875 North Michigan Avenue<br>Suite 3434<br>Chicago, IL 60611-1901<br>phone: (312) 266-8680 | Bonnie Krueger<br>Editor in Chief<br>Martha Carlson<br>Associate Editor | |

Titles include *Sophisticate's Hairstyle Guide*, and more.

| | | |
|---|---|---|
| *Sophisticate's Black*<br>*Hairstyles Guide*<br>875 North Michigan Avenue<br>Suite 3434<br>Chicago, IL 60611-1901<br>phone: (312) 266-8680 | Jocelyn Amador<br>Associate Editor | |

# Appendix 4

# Trade Magazines

Getting published in your own industry can lead to many other PR opportunities, as discussed throughout this guide. Targeting these magazines can pay off in many big ways for you.

Following are the names, addresses, phone number and contacts for major trade magazines covering hair, skin, nails, and tanning in the professional salon industry.

Keep in mind that staff changes occur almost monthly. Magazines can also move, be sold, or cease publication altogether. Therefore, it's important that you call to verify the contact name, address, telephone number, and fax number before sending press materials to the publication. Once you launch your PR program, continue to verify and update your list quarterly.

Start by calling these magazines to ask for an editorial calendar. This is a monthly breakdown on the focal points for each issue, such as perms, hair color, holiday hair, nails, retailing, etc. The editorial calendars will be useful to you when you're putting together your own PR program. Keep in mind that editors are working on these topics 4 to 6 months in advance of their publication date.

To add to your list, when you attend trade shows, keep an eye out for new magazines and collect all pertinent editorial contact and subscription information. If you're lucky, you might even get to meet an editor or writer at the booth.

Finally, make sure you subscribe to all of the major magazines, as well as to any of the more specific titles, such as nails and skin care, that apply to your business. You'll want to make sure that you're up monthly on the kind of information each publication runs. You'll also want to keep an eye on how other salons are getting published.

## Trade Magazines

| PUBLICATION<br>ADDRESS<br>PHONE/FAX | CONTACT<br>NAME/<br>TITLE | MY NOTES |
|---|---|---|
| *American Salon*<br>270 Madison Avenue<br>New York, NY 10016<br>phone: (212) 951-6600<br>fax: (212) 481-6562 | Lorraine Korman<br>Editor | Full service |
| *Modern Salon* (main office)<br>400 Knightsbridge Parkway<br>Lincolnshire, IL 60069<br>phone: (708) 634-2600<br>fax: (708) 634-4379 | Jackie Summers<br>Editor in Chief<br><br>Arlene Tolin<br>Creative Director | Full service |
| *Modern Salon* (East Coast)<br>370 Lexington Avenue<br>Suite 2001<br>New York, NY 10017<br>phone: (212) 682-7777<br>fax: (212) 682-7562 | Maggie Mulhern<br>Beauty Editor | |
| *Modern Salon* (West Coast)<br>6800 Owensmouth<br>Suite 430<br>Canoga Park, CA 91303-2091<br>phone: (818) 716-5800<br>fax: (818) 716-6019 | Lisa Goldman<br>West Coast Editor | |
| *Salon Today*<br>400 Knightsbridge Parkway<br>Lincolnshire, IL 60069<br>phone: (708) 634-2600<br>fax: (708) 634-4379 | Michele Musgrove<br>Editor | Salon management |
| *Salon News*<br>Fairchild Publications<br>7 West 34th Street<br>New York, NY 10001<br>phone: (212) 630-3547<br>fax: (212) 630-4511 | Melissa Bedolis<br>Editor | Full service |
| *SalonOvations*<br>P.O. Box 98<br>El Paso, IL 61738<br>phone/fax: (309) 527-5060 | Barbara Jewett<br>Managing Editor | |

| PUBLICATION ADDRESS PHONE/FAX | CONTACT NAME/ TITLE | MY NOTES |
|---|---|---|
| *Hair & Beauty News* #302 The Cezanne 712 W. 48th Street Kansas City, MO 64112 phone/fax: (816) 756-3336 | Marty McCarty Editor | |
| Victoria Wurdinger 96 Sterling Place Suite 2B Brooklyn, NY 11217 phone/fax: (718) 857-6483 | | freelance writer |
| Stephanie Pedersen 327 E. 34th Street, #4C New York, NY 10016 phone: (212) 725-6605 fax: (212) 725-6607 | | freelance writer |
| *Hair International News* 124-B East Main Street PO Box 273 Palmyra, PA 17078 phone: (717) 838-0795 fax: (717) 838-0796 | | (PA and NJ) |
| *California Stylist/ Washington Stylist* 6700 SW 105th, Suite 309 Beaverton, OR 97005 phone/fax: (503) 646-5646 | Linda Holland Editor | (Northwest region) |
| *Chic* (Chicago Cosmetologists Association) 401 North Michigan Avenue Chicago, IL 60611 phone: (312) 644-6610 fax: (312) 321-6889 | Patrick O'Keefe | Chicago area |
| *Canadian Hairdresser* 132 Cumberland Street 3rd Floor Toronto, Ontario, Canada M5R186 phone: (416) 923-1111 fax: (416) 964-1031 | Joan Harrison Editor | Canada |

| PUBLICATION<br>ADDRESS<br>PHONE/FAX | CONTACT<br>NAME/<br>TITLE | MY NOTES |
|---|---|---|
| *Salon Magazine*<br>411 Richmond Street East<br>Suite 300<br>Toronto, Ontario Canada M5A3S5<br>phone: (416) 869-3131<br>fax: (416) 869-3008 | Alison Wood<br>Editor | Canada |
| *Styliste Coiffure*<br>Novostyl International<br>1001 Notre Dame<br>Repentigny<br>Quebec, Canada J5Y 1E1<br>phone: (514) 581-7230 | Guy Boucher | Canada<br><br>(biannually, Sept.<br>and March), send<br>b&w photos |
| *Snippets*<br>Suite 210<br>1755 West Broadway Avenue<br>Vancouver, BC V6J 4S5<br>phone: (604) 736-9891<br>fax: (604) 736-0720 | Andrea Sinclair | Canada |
| *"Hair" the News*<br>54 Pottery Crescent<br>Brampton, Ontario Canada L6S 3S3<br>phone: (905) 454-0112 | Denise Thompson<br>Editor | Canada<br>(Ethnic trade) |
| *Passion*<br>*Black Passion*<br>*Coiffure Q*<br>Dowa Planning, Inc.<br>Dairoku Seiko Building<br>1-31 Akasaka, 5-Chome<br>Minato-ku, Tokyo 107 Japan<br>phone: 03-3589-2571<br>fax: 03-3589-2802 | Helen Moy<br>Editor | International<br>Styles selectors |
| *MOODS*<br>1249 S. Diamond Bar Blvd., #316<br>Diamond Bar, CA 91765<br>phone: (714) 597-1455<br>fax: (714) 597-7169 | | African American |

| PUBLICATION<br>ADDRESS<br>PHONE/FAX | CONTACT<br>NAME/<br>TITLE | MY NOTES |
|---|---|---|
| *Intra-America Beauty Network*<br>14 Commerce Drive, P.O. Box 629<br>North Branford, CT 06471<br>phone: (800) 634-8500<br>call for guidelines and issue themes | | Styles selector |
| *Les Nouvelles Esthetiques*<br>306 Alcazar Avenue #204<br>Coral Gables, FL 33134<br>phone: (305) 443-2322<br>fax: (305) 443-1664 | Ruth Zelouf Elias<br>Editor | Skin care |
| *Skin Inc.*<br>P.O. Box 318<br>Wheaton, IL 60189<br>phone: (708) 653-2155<br>fax: (708) 653-2192 | Marian Raney<br>Editor | Skin care<br>(business) |
| *Dermascope*<br>3939 East Highway 80<br>Suite #408<br>Mesquite, TX 75150<br>phone: (214) 682-9510<br>fax: (214) 686-5901 | Naomi Stokes-Wesson<br>Editor | Skin care |
| *Day Spa*<br>7628 Densmone Ave.<br>Van Nuys, CA 91406-2088<br>phone: (800) 422-5667<br>fax: (818) 782-7328 | Linda Lewis<br>Editor | Day Spa |
| *Tanning Trends*<br>P.O. Box 1630<br>Jackson, MI 49204<br>phone: (517) 784-1772<br>fax: (517) 787-3940 | Kristy Miller<br>Editor | Tanning |
| *Today's Image*<br>8888 Thorne Road<br>Horton, MI 49246<br>phone: (517) 563-8133<br>fax: (517) 563-8833 | John Dancer | Tanning |

## 162  Trade Magazines

| PUBLICATION<br>ADDRESS<br>PHONE/FAX | CONTACT<br>NAME/<br>TITLE | MY NOTES |
|---|---|---|
| *Looking Fit*<br>4141 North Scottsdale Road<br>Suite 316<br>Scottsdale, AZ 85251<br>phone: (602) 483-0014<br>fax: (602) 990-0819 | John Lyren | Tanning |
| *Nails Magazine*<br>2512 Artesia Blvd.<br>Redondo Beach, CA 90278<br>phone: (310) 376-8788<br>fax: (310) 376-9043 | Cyndy Drummey<br>Editor in Chief | Nails |
| *Nail Pro*<br>7628 Densmore Avenue<br>Van Nuys, CA 91406-2088<br>phone: (800) 422-5667<br>fax: (818) 782-7328 | Linda Lewis<br>Editor | Nails |
| *BBSI's Beauty Inc.*<br>11811 N. Tatum Blvd.<br>Suite 1085<br>Phoenix, AZ 85028<br>phone: (602) 404-1800<br>fax: (602) 404-8900 | Denise M. Rucci<br>Editor | Distributor<br>publication |
| *Beauty Store Business*<br>9600 Lurline Avenue<br>Chatsworth, CA 91311<br>phone: (818) 998-1811<br>fax: (818) 341-5298 | Lou Greco<br>Editor | Beauty stores |
| *Salon Business Strategies*<br>PO Box 296, 40 Main Street<br>Centerbrook, CT 06409<br>phone: (203) 767-2064<br>fax: (203) 767-2084 | Debra A. Majewicz<br>Editor | Business |
| *HAIRNET*<br>383 Grand Street<br>New York, NY 10002<br>(212) 505-1936<br>http://www.hairnet.com | Larry Tashman | InterNet |

# Appendix 5

# PR Activity Log

Use these pages to keep track of all conversations you have with media contacts.

| Date | Name of Media | Contact Names | Phone # | Topic of Conversation |
|------|---------------|---------------|---------|-----------------------|
|      |               |               |         |                       |

## PR Activity Log

| Date | Name of Media | Contact Names | Phone # | Topic of Conversation |
|------|---------------|---------------|---------|-----------------------|
|      |               |               |         |                       |

# Appendix 6

# Press Clippings Record

Use these pages to keep track of your PR successes, including mentions in newspapers, magazines, on TV and radio, photo credits, appearances, etc.

| Media | Issue | Type of Coverage | Name of Contact |
|---|---|---|---|
| | | | |

## Press Clippings Record

| Media | Issue | Type of Coverage | Name of Contact |
|-------|-------|------------------|-----------------|

# Appendix 7

# Press Contacts

Add your own press contacts here:

| PUBLICATION<br>ADDRESS<br>PHONE/FAX | CONTACT<br>NAME/<br>TITLE | MY NOTES |
|---|---|---|
| | | |

## Press Contacts

| PUBLICATION<br>ADDRESS<br>PHONE/FAX | CONTACT<br>NAME/<br>TITLE | MY NOTES |
| --- | --- | --- |

# Glossary/Index

Note: 1-Page numbers in **bold type** reference non-text material.
2-Titles and definitions are in **bold type.**

## A
**Advertising is paid publicity,** 7
Agent
   budgeting for, 64-71
      form, **65**
   choosing, 63
   need for, 59-60
   screening, 61-63
**All About You,** 150
**Allure,** 150
Alvarez, Frank, on public relations, **81**
**American Salon,** 158
   editors' desires at, 106-9
Art auction, as PR event, 124
**Art is the slides, photos, or illustrations that can be used to accompany your idea,** 11

## B
Back to school fashion show, as PR event, 125
**Backgrounder is a complete press kit that tells everything possible about you, your salon, and your staff that's relevant to your PR program,** 35-37
**Barter refers to trading services,** 67
BBSI's Beauty Inc., 162
Beauty parties, 122
**Beauty Store Business,** 162
**Bio includes biographies of you and any of your top staff members who will be featured in your PR,** 35
**Black Elegance,** 150
**Black Hair Care,** 156
**Black Hairstyles,** 156
**Black Passion,** 160
**Black-Tress,** 156
Bridal beauty expo & champagne party, as PR event, 125
Bridal fairs, 121

**Bridal Guide,** 150
**Brides,** 150
Briggs, Jesse, PR strategies of, 20-22
Budgeting
    form, **65**
        local event checklist and, 75
        for professional writer, 64, 66-68
        for public relations agent, 64-71
Business card, 29, 30

## C

Calsacola, Richard, on charity work, 124-25
**Canadian Hairdresser,** 159
**Celebrity Hairstyles,** 155
    editors' desires at, 91-92
Champagne & bridal beauty expo, as PR event, 125
**Charity work is also known as cause marketing and involves raising money, collecting food or clothing, or calling attention to a cause that's important to a salon's staff and clients,** 123-29
    Richard Calsacola on, 124-25
    Robert Lamorte on, 126-27
**Chic,** 160
**Clippings are the editorial mentions you get as a result of your PR efforts,** 13
**Coiffure Q,** 160
    editors' desires at, 115-17
**Complete Hair and Beauty Guide,** 156
    editors' desires at, 92-93
**Complete Short Hair Styling Guide,** 156
    editors' desires at, 92-93
Consistency, success and, 8, 9
**Consumer magazines are those publications that are read by the general public,** 13
    list of, 149-53
    public relations program aimed toward, 46
**Cosmopolitan,** 150
Cut- & color-a-thon, as PR event, 124

## D

**Dermascope,** 161
Designer, budgeting for, 68
Designers, 25
Discounts, press and, 78

## E

Editors, contacting, 86-87
ELLE, 152
Essence, 152
The Exclusive is a story that you offer to one publication only and do not release anywhere else until the first publication refuses to use it, 10, 11

## F

Face, 151
Fact sheet is a brief press release describing your salon and what makes it unique, 35
Family Circle, 151
Fashion show, as PR event, 124
Feature release is a "tip" release or a "story," 37
First for Women, 151
Flyers, 29, **30**
Fontanez, Edwin, on public relations, **79-80**
Forms
   budget, local event checklist and, 75
   Model Release Form, 33
   Photographer's Release Form, 34
Free health workshops, as PR event, 124-25
**Freelance writers work independently and are hired by a variety of publications to write articles and research various topics, 11**
   editors' desires and, **69-71**

## G

GCR Publishing
   editors' desires at, 92-93
   hair style magazines and, 156
**Glamour**, 151
   public relations program aimed toward, 46
Goals, of public relations program, 43-44
**Good Housekeeping**, 151
   editors' desires at, 89

## H

**Hair & Beauty News**, 159
**Hair International News**, 159

"Hair" the News, 160
HairDo, 155
**HairDo Ideas,** editors' desires at, 91-92
**Hairdo and Makeover,** editors' desires at, 92-93
Hairmeister of Las Olas, image of, 20-22
**HAIRNET,** 165
Hairstyling publications, 91-93, 155-56
**Harper's Bazaar,** 151
Harris publications, 155-56
   editors' desires at, 91-92
Health workshops, as PR event, 124-25

I

Image
   backgrounder and, 35-37
   collateral materials, 25-30
   enhancing, 129-31
      ideas for, 129-41
   logo and, 26-27
   model releases and, 10, 32-35
   photo shoots and, 30-32
   press,
      folders, 28-29
      releases and, 37-40
   relationship building and, 40-41
   salon's, 19-20
      Hairmeister of Las Olas, 20-22
   stationary/letterhead and, 27
   worksheet, 22-25
Industry, public relations program, 46-49
**Instant Hair Styles,** 156
   editors' desires at, 92-93
Intercoiffure, joining, 118-19
**Intra-America Beauty Network,** 161

L

LaCour, Lenny, on public relations, 87-88
**Ladies Home Journal,** 151
LaMorte, Robert, on charity work, 126-27
**Lead time is how far in advance an editor is working on an issue,** 12
**Les Nouvelle Esthetiques, 48,** 161
Letterhead, selecting, 27

List, local media, 143-48
Local media list, 143-48
Local public relations program, 45, 73-74
   beginning a, 76-84
   charity work, 123-29
   Edwin Fontanez on, **79-80**
   event checklist/budget form, 75
   fashion show, 125
   Frank Alvarez on, **81**
   health workshop, 124-25
   press discount policy, 78
   receptions, 122
   reasons for, 75-76
   silent art auction, 124
   special events, 121-23
Logo, selecting, 26-27
**Looking Fit**, 162

## M

**Mademoiselle**, 151
**Marie Claire**, 152
   editors' desires at, 90
**The Maskhead is the column or page in the magazine that lists the names of the editorial staff and often freelance writers**, 12
Materials
  image and, 25-30
    logo, 26-27
    press folders, 28-29
    stationary/letterhead, 27
**McCall's**, 152
Media
  list, local, 143-48
  salons and, 1-2
**Menu is your salon's services and prices**, 35
**Mirabella**, 152
Model Release Form, 33
**Model releases are legal documents that are signed by the models giving you the right to use their image in your press**, 10, 32-35
**Modern Bride**, 152
**Modern Salon**, 158
  editors' desires at, 99-104
  public relations program aimed toward, 46

**Monthly retainer is a set fee per month and gives you the greatest consistency and an assurance that someone is always looking out for your press needs and ensures more consistent follow-up,** 66
**MOODS,** 160
Motivation, staff, public relations and, 4-5

## N

Nail Industry Association, joining, 118-19
**Nail Pro,** 162
**Nails Magazine, 48,** 162
   editors' desires at, 112-14
National Cosmetology Association, joining, 118-19
National public relations programs, 46, 85-94
   editors desires and, 89-93
   Lenny LaCour on, 87-88
   television and, 94-95
National publications, 89-91
Networking relationships, building, 40-41
**101 New Hairstyle Ideas,** editors' desires at, 92-93
**New Ideas for Hairstyling,** 155
**New Woman,** 152
News, public relations and, 3
**News release is something that's current. It is usually time specific, such as a cut-a-thon on April,** 12, 37

## P

**Passion,** 160
   editors' desires at, 115-17
**Pay as you go is paying for a particular project,** 67
Pedersen, Stephanie, 159
**Per project is pay as you go on a particular project,** 67
Photo shoots, 68
   image and, 30-32
**Photographer releases are documents signed by photographers that give you the right to use their work in your press,** 10, 32-35
Photographer's Release Form, 34
**Picture This....Your Image In Print,** 31
**The Pitch is how you present your idea to an editor,** 11
Press
   clippings record, 165-66
   contacts, 12, 167-68
   discount policy, 78

## Glossary/Index 175

folders, image and, 28-29
releases, 37
**Press book is the book in which you keep your clippings,** 13, 165-69
**Press list is your current list of editors, writers, and television or radio segment producers, plus all contact information and notes on interactions you had with them,** 12, 167-68
Printing, budgeting for, 68
Professional writer, budgeting for, 64, 66-68
Public Relations Activity Log, 163-64
Public relations agent, 8
    budgeting for, 64-71
        form, 65
    choosing, 63
    need for, 59-60
    screening, 61-63
**Public relations is the art and business of shaping the public's opinion about you and your business,** 7
    consistency and, 8, 9
    goals of, 43-44
    groups, 118-19
    ideas for, 131-41
    maximizing, 117-18
    rules for successful, 15-18
    strategies of, Jesse Briggs, 20-22
**Public relations program is the process by which you generate editorial and good will throughout your community through your local media and activities and to other target audiences,** 2
    benefits of, 2-5
    developing your, 44-45
    enhancing, 129-31
        ideas for, 131-41
    example, 50-55, 55-58
    goals of, 43-44
    local,
        beginning, 76-84
        charity work, 123-29
        Edwin Fontanez on, **79-80**
        events checklist/budget form, 75
        fashion show, 125
        health workshop, 124-25
        press discount policy, 78
        receptions, 122

    reasons for, 75-76
    silent art action, 124
    special events, 121-23
  outlining your, 55-57
  results of, 3-4
  targeting, 57
    message of, 45-49

# R
Receptions, as public relations events, 122
**Redbook**, 152
  editors' desires at, 90-91
Relationships, building, 40-41
Releases, model/photographer, 10, 32-35

# S
Salon
  anniversaries as PR, 123
  grand openings of, 121
  image of, 19-20
    Hairmeister of Las Olas, 20-22
    **menu includes your salon's services and prices**, 35
  image and, 29, 30
Salon Association, joining, 118-19
**Salon Business Strategies**, 162
**Salon Magazine**, 160
**Salon News**, 158
  editors' desires at, 105-6
  public relations program aimed toward, 46
**Salon Today**, 158
**SalonOvations**, 158
  editors' desires at, 109-11
  public relations program aimed toward, 46
**Segment is an all-inclusive of a TV or radio program**, 12
**Self**, 152
  editors' desires at, 89
**Seventeen**, 152
**Short Hair**, editors' desires at, 91-92
**Short Hair Styles**, 155, 156
  editors' desires at, 92-93
Silent art auction, as PR event, 124

Glossary/Index **177**

Skin Inc., 161
Snippets, 160
Soap Star Hair Styles, 155
   editors' desires at, 91-92
**Sophisticate's Black Hairstyle Guide**, editors' desires at, 93, **94**, 156
**Sophisticate's Hairstyle Guide**, editors' desires at, 93, **94**, 156
Special events, 121-23
Staff motivation, public relations and, 4-5
Stationary, selecting, 27
**Step-By-Step Hairdo Ideas**, 158
**Step-by-Step Hairstyling**, editors' desires at, 92-93
**Styliste Coiffure**, 160


T
**Tanning Trends**, 161
Teen, 150, 152
Television, national, 94-95
Third-party endorsement, someone else speaking well of you, 2
Today's Image, 162
**Trade magazines are read by other beauty professionals**, 14, 157-62
   editors' desires at,
      American Salon, 106-9
      Coiffure Q, 115-17
      Modern Salon, 99-104
      Nails magazine, 112-14
      Passion, 115-17
      Salon News, 105-6
      SalonOvations, 109-11
   public relations program aimed toward, 46-49, 97
Travel, budgeting for, 68
**Trend release is your prediction for hair, beauty, and fashion trends for the upcoming season**, 37
   sample of, 40


V
Vogue, 152

W
W, 153
Woman's Day, 153
   public relations program aimed toward, 46

**Woman's World**, 153
Worksheet, image, 22-25
Writer, budgeting for, 64, 66-68
Wurdinger, Victoria, 159
    **Picture This....Your Image In Print**, 31

# Y
Yellow Strawberry Global Salon, 20
**YM**, 153